Making Machines of Animals

Animals, History, Culture

HARRIET RITVO, SERIES EDITOR

MAKING MACHINES OF ANIMALS

The International Livestock Exposition

Neal A. Knapp

Johns Hopkins University Press

Baltimore

This book was brought to publication through the generous support of the Estate of Mr. J. R. Herbert Boone endowment, with a matching gift from the National Endowment for the Humanities.

Johns Hopkins University Press
2715 North Charles Street
Baltimore, Maryland 21218
www.press.jhu.edu

Cataloging-in-Publication Data is available from the Library of Congress.

A catalog record for this book is available from the British Library.

ISBN: 978-1-4214-4655-4 (hardcover)
ISBN: 978-1-4214-4656-1 (ebook)

Special discounts are available for bulk purchases of this book. For more information, please contact Special Sales at specialsales@jh.edu.

For Adrea

CONTENTS

AHSP	Alvin Howard Sanders Papers, Division of Rare Manuscript Collections, Cornell University Library, Ithaca, NY
CADAS	College of Agriculture, Department of Animal Sciences records, addition 2, Purdue University, West Lafayette, IN
CFCP	Charles F. Curtiss Papers, RS 9/1/12, Special Collections Department, Iowa State University Library, Ames, IA
CHMRC	Chicago History Museum Research Center, Chicago
CSC	Chicago Stockyards Collection. Special Collections and University Archives, University of Illinois at Chicago
CSPP-OSU	Charles Sumner Plumb Papers, Ohio State University Archives, Columbus, OH
CSPP-PU	Charles Sumner Plumb Papers, Purdue University Archives and Special Collections, West Lafayette, IN
GICP	G. I. Christie Papers, Purdue University Archives and Special Collections, West Lafayette, IN
ILER	International Livestock Exposition Records. Special Collections Department, Iowa State University Library, Ames, IA
JHSP	John H. Shepperd Papers, North Dakota State University Libraries, Fargo, ND
LHBP	Liberty Hyde Bailey Papers, Division of Rare Manuscript Collections, Cornell University Library, Ithaca, NY
PSSP	P. S. Shearer Papers, Special Collections Department, Iowa State University Library, Ames, IA
SASC	Saddle and Sirloin Club publications, North Dakota State University Libraries, Fargo, ND

Making Machines of Animals

The World's Most Conspicuous
Livestock Show

> The mission of the International Live Stock Exposition . . . was to gather
> into one place the best specimens of cattle, sheep, swine and horses that
> could be found, and thereby present to the agricultural population of the
> United States a great and valuable educational opportunity, wherein the
> eye and the mind should be instructed and encouraged to the production
> of better animals for breeding, marketing and exporting.
>
> A Review of the International Live Stock Exposition, 1913

The Union Stock Yard and Transit Company in Chicago hosted the first International Livestock Exposition in 1900. When animals arrived for the show, handlers immediately directed them down chutes and through gates to their pens, a "hotel" for livestock. To reach the exhibition halls, exhibitors and spectators entered through the stockyard's three-arched gate built in 1879 and designed by architects Daniel Burnham and John Root (fig. I.1). The gate's middle arch featured a sculpture of the head of a bull named Sherman—a bull that won the grand sweepstakes at the first American Fat Stock Show in 1878. Of course, the slaughter animals that entered the stockyards faced different fates than did the International show animals and prize-winning bulls like Sherman.[1]

From 1900 to 1975, the International awarded premiums, special prizes, trophies, and badges, drawing over ten thousand animals each year. As the associate editor of the Breeder's Gazette exclaimed following the inaugural show, the exposition dwarfed "all shows of recent years" and its "wonders . . . [were] unrivaled in history."[2] Farmers and urban spectators flocked to it; in 1904, 1905, and 1906, the average attendance per year reached four hundred thousand during the December show week. Railroad companies offered reduced fees to encourage people to attend. Not only did attendees fill the amphitheater to capacity, but Chicago's boardinghouses, hotels, streetcars, and restaurants overflowed with both American and foreign visitors. These guests traveled to the stockyards to view the many animals and competitions, which included agricultural college exhibits and demonstrations by the federal government.[3]

Figure I.1. Three-arched gate built in 1879 for the Union Stockyards in Chicago. *Source: A Review of the International Live Stock Exposition,* 1916.

The International brought together livestock producers, land-grant college researchers, and government officials. But it represented much more to these visitors than an excursion to Chicago and some spectator fun. The International operated at the center of a broader web of institutions and reformers in the national livestock improvement movement devoted to transforming husbandry practices as well as the animal body. The cattle, sheep, and pigs unloaded from the trains in Chicago represented works in progress. The International organizers and judges used a system of incentives and penalties to reconfigure genetic makeup and physical shape. In the show ring, judges selected and normalized what they considered to be modern bodies and physiological types that correlated with farm function—or commodity specialty like meat or dairy—a turn in animal husbandry that reflected key components of industrial production: specialization and standardization. In doing so, the exposition encouraged farmers to effect changes in their practices to meet the demands of the meat trust in Chicago and to adopt modern methods of food production.[4] The exposition thus employed spectacle to disseminate industrial ideologies that unambiguously applied to the animal body.

By the turn of the century, Chicago had emerged as the industrial center of the agricultural United States as meatpackers worked to vertically consolidate the industry.[5] They bought, processed, and shipped livestock and animal-based products at the Union Stockyards, controlling every aspect of the meat business from the purchasing and slaughtering of the animals to sales at the retail counter. They faced the wrath of "muckraking" journalists, distrustful consumers, and angry producers, who opposed the consolidation of the industry and leveled price-fixing and food quality charges against the packers.[6] And indeed the International Livestock Exposition was meant to function in part as a "pure food display" that would dispel the consumer distrust provoked by unsettling events such as the embalmed meat crisis during the Spanish-American War, which resulted in the deaths of soldiers, and a rising storm of negative publicity culminating with the 1906 publication of Upton Sinclair's *The Jungle*. Beautifully groomed animals might recast the industry as an enterprise aiming for a more humane pursuit of healthy, safe food. But the meatpackers' primary motive, and the main reason for the industry's financial and institutional support of the International, was a more pressing concern over the quality and quantity of livestock available to them.

Despite the acknowledged power of the Chicago meatpacking industry at the close of the nineteenth century, the packers felt genuine supply-side vulnerabilities and fretted about limited control over animal quality, uniformity, and timing to market. The limited supply of beef in the 1890s worried the packers. As the demand for meat grew in America's cities, the number of meat-producing animals available declined. Packers blamed suppliers and farmers for price fluctuations, while their critics believed, of course, that the packers' control over the industry allowed them to inflate prices.[7] This supply problem, however, had as much to do with packers' industrial successes as with farmer behavior and animal bodies. In vertically consolidating the meat industry, the packers thrust the vagaries and uncertainties of life onto farmers and ranchers. The packers' business model was fundamentally rooted in handling animals only momentarily in the food production sequence. The farmers and feeders bore the risks of nature and animal biology, leaving the seemingly all-powerful packers ill suited to structurally effect change in animal supply. Packers thus insulated themselves from the risks associated with the variability of life—a central characteristic of the packer profitability—leaving those problems, including economic insecurity, for livestock producers to manage. However, in eliminating life as an obstacle in their profit-making model, the packers also were left vulnerable to a persistent, perennial supply problem that owed to their limited control over the feeding, caring for, and marketing of live animals from birth to finish. The International gave the packers leverage over animal life and the opportunity to remake these biological beings in the industrial image.

Meatpackers provided the financial foundation for the International, but it could not have functioned without a second set of participants: a deeply invested group of land-grant university researchers who worried about the precarious relationship between national agricultural output and rural stability. These economists and crop and animal husbandry experts wanted farm productivity and food production to keep pace with the explosive growth of urban populations, even as farm areas themselves lost population and fewer farmers remained tied to the countryside. They also wanted rural standards of living to more closely match urban standards and hoped such improvements would entice rural residents to stay on the farm. Increased incomes required increases in output, which would also address the nation's growing food demands.[8]

Increasing both farm income and farm stability necessitated major shifts in agricultural practice. With new lands no longer available as the frontier closed, these scientists and social scientists feared diminished farm output. To resolve this problem, the professors pushed "permanent farming" or "balanced farming," which required the use of livestock on every farm. Animal waste replenished the soil with nutrients and organic matter. Not only should every farm keep livestock but agricultural improvement obliged farmers to breed better stock—animals that efficiently converted grain, not wild forage, to human food. Raising livestock specialized in meat production allowed farmers to feed corn to animals and then sell surplus meat for a profit.

What bound together these two unlikely interests—Chicago meatpackers and land-grant university professors—was a focus on the transformation of meat-producing animals. The variability, the uncertainty, and the risks associated with animal life became the focus of these two groups' work and the site of the intersection of their different agendas. Animal life was the central problem to be solved in this effort to modernize and industrialize American agriculture. The packers and professors thus found mutual benefit in livestock improvement and hoped transforming livestock could solve the supply problem, increase food production and thereby farmer incomes, and contribute to maintaining soil fertility for the greater good of American land. The International promoted clear guidelines for selecting and raising these animals. Among the most important tenets was the need to feed grain to livestock, as doing so regulated both meat quality and the timing of slaughter. Grain hastened the pace of growth, muscle development, and fat cover. By using stored grain and roughages like hay, feeders could fatten animals and send them to the market more steadily throughout the year, which helped ameliorate the seasonality of supply. The stockyards' industrial model required constant access to meat-producing animals, and feed and feedlot husbandry reoriented the sequencing of production and mitigated the impact weather and growing seasons had on the flow of animals to the market.

The professors had a different conception of time from the packers. In an effort to improve the farmer's monetary return, these researchers advocated ideas of efficiency that obliged producers to consider input cost in relation to the sale value of livestock. The most significant input consideration in grain-fed livestock production was time. Each passing day cost the farmer more money to grow animals, and every pound after an animal's birth cost slightly more to add than the day before. Thus, weight gain cost more as livestock aged; furthermore, meat tenderness and quality diminished with older age. The economists and scientists applied a rates-based model to feed conversion, nutritional waste, and age-to-finish costs and compared these input variables to market value. They urged farmers to disregard the gross value of animals and focus instead on the marginal returns gleaned from livestock production. Simply put, the heaviest animals sold on the market were no longer the most profitable.

But grain alone could not transform animal husbandry into a modern enterprise. Progressive breeding also necessitated—according to the packers and professors at the International—the injection and propagation of purebred British genetics in American livestock, which they agreed required the extermination of inferior animals, or so-called scrubs, through castration, selective breeding, and killing. Improved livestock with a proper genetic makeup provided quicker and larger returns for the feed consumed by the animal. The animals' bodies, like standard products assembled in the factory, needed to be uniform, efficient in design, and oriented toward the production of one commodity.[9] Preferences shifted in favor of purebred genetics and small, early maturing cattle, sheep, and pigs that were efficient and higher yielding. The packers benefited from product consistency, but the smaller and more compact bodies also helped them to fit more edible meat on a railcar during live shipment and to extract more valuable meat per animal on the disassembly lines in Chicago.

The mutual interests of Chicago packers and land-grant professors led them to collaborate to create the International Livestock Exposition as the hub for the livestock improvement movement. The International's early success also led to the US Department of Agriculture's endorsement and direct participation, as well as the formation and hosting of issue-oriented associations to foster scientific relationships and to encourage and disseminate agricultural research. These professional associations supported the International's central goals and provided the International with organizational backup in research areas like animal nutrition, genetics and breeding, and scientific grain production, which directly linked farmers to an emerging class of nonfarm agricultural salespeople and experts who connected these research categories associated with scientific farming to commercial inputs.

The packers' industrial ideology served the interests of processors and middlemen—those who produce and peddle capital intensive inputs—creating a national complex of agribusiness specialists who neither worked on the farm nor produced animals or grain, and that ideology continues to serve those same interests today. Animal standardization and specialization undermined local context-dependent knowledge of different regions and ecologies. Instead of addressing soil exhaustion inherent to the extensive farming models of the nineteenth century, industrial farming mirrored and relied on the extractive, surplus-oriented economic models followed by mines and factories.

Soil fertility was not addressed by balanced or permanent farming, but nonfarm agricultural specialists and their clients, the farmers, relied on the redistribution of finite resources to maintain or increase yield. After harvesting Corn Belt grains, farmers had to sell their products for the animals to work as cogs in this sequence that gave value to the grains as well as to all the machines and inputs to create them by transmuting corn to meat for human consumption. The modern animal required farm specialization and industrialization, but at the same time the new-age, capitalistic farmer and all the middlemen needed the modern animal to do the work of creating surplus value from land, synthetic fertilizers, commercial seeds, and grains. The International promoted industrial ideologies that, when put in practice on the farm, spurred producer dependency on an emerging class of nonfarm agricultural specialists that simultaneously promoted and served the interests of scientific crop and animal husbandry.

As the apex of this improvement movement under which professional organizations, land-grant experts, and nonfarm agricultural specialists organized, the International leveraged the animal fantasy—the remaking of animal bodies magnified by a spectacle on the scale of a world's fair—to mechanize the food production sequence. The International was therefore not simply an event but an institution that through incentives and penalties denormalized and renormalized modern biological beings and a complex, organizational ecosystem that propagated this revolution in American agriculture. For example, the stockyards housed and underwrote purebred associations, which provided the organizational framework for competition. The International's classes were structured around breed. The use of a purebred animal of British descent became a prerequisite for participation. Even in events aimed at improving commercial breeding, like the mutton competition, participants were required to breed average females of unknown genetics to purebred sires to demonstrate the value of well-bred sires to college students and livestock breeders.

Their known ancestries and uniform phenotypical traits provided information to breeders about probable productivity in future matings. But purebreds had limits; these well-bred animals were not considered fixed, permanent, or already complete. Genetics alone did not constitute the modern animal. Instead, purebred animals provided only the foundation for the "ideal" animal. The International also focused on the physical makeup of meat-producing livestock beyond the aesthetic or "fancy" traits, like color, ear shape, or horns, often associated with purebred phenotypes. In this way, the International set out to alter animal bodies, and the show ring dictated these forms. Showmen led cattle with a halter and sheep by hand, and they guided pigs with a nicely fashioned stick or cane around the ring for evaluation. The judges looked at animal depth, width, and thickness. They "handled" the most valuable parts, like the rack, loin, and leg on the sheep, to assess and compare the estimated market value of each animal. Finally, the judges placed the animals based on these visual and tactile metrics.

Meat-producing cattle preferred by the exposition's judges replaced the old, tall, and thin nineteenth-century range cattle. The modern meat animal was compact with a rectangular, bulky body. Shorter in stature and younger in age, less than twenty-four-months old, the ideal market steer produced a higher-yielding carcass, not a larger overall carcass weight. The prioritizing of these traits led showmen to utterly reject older steers, and so aged market animals all but vanished in just the first two decades of the show. New-age livestock with specific forms and functions depended on specific farm structures—high input costs and capital- and labor-intensive farm strategies. The packers and professors thus leveraged the International to industrialize animal life itself and spur specialization and standardization on the American farm.

This book documents the various challenges, ills, and the benefits of the national livestock improvement movement by looking at the central influence of this Chicago exposition that the packers and professors created. The International promulgated and promoted industrial standards for the animal body and the modern farm, serving as both the standard maker and the final arbiter for the livestock world, settling disputes within the agricultural community and resolving disparities in judgment at county and state fairs. Its founders often touted the show's central importance by comparing it to the US Supreme Court. Even though the show could not legally enforce its standards, it created norms by shaping breeders' tastes and preferences and by penalizing farmers and animals who did not toe the line with disqualification or low placings. In its role as hub, the International established the criteria for animal evaluation, and it influenced the advice and information distributed by land-grant affiliated agencies that directly interacted with farmers. Its standards therefore radiated out into the agricultural community. To win at state fairs around the country and receive positive reviews from judges, exhibitors bred their animals in a way that was geared toward achieving the International's goals.

★

Spectators from around the world and progressive farmers from every nook of American life traveled to the International to witness and participate in the marvel that was the so-called eighth wonder of the world. In fact, the throngs of people that flocked to Chicago by boat, train, horse, and eventually motorized vehicle were as much a part of the delectation as the state-of-the-art buildings and the domesticate broods. The International's pageantry unveiled intricate aspects of the evolution of American life in the twentieth century. Urban and suburban consumers across the United States grew increasingly detached from the animal experience. Industrialization and off-the-farm food exchanges distanced consumers from the sources of the food on their tables.

Technological changes, including the motorized vehicle, and municipal policies that limited the presence of livestock in urban spaces exacerbated the separation between animals and humans that characterized modern life in emerging metropolises.[10]

One of the most important pieces of emerging technologies was the animal itself. The International incentivized the development of biotic technology that aggravated the physical and emotional void between human consumers and the killed and disassembled animal. But farmers too participated in this spatial and emotive distancing. The life experiences of farm human and farm animal, as well as their interactions, were overhauled owing to the resequencing of production and the drive for profits. The efficiency-oriented goals of the reformers directed this development.

Engineers and scientists used efficiency as a standard to measure the performance of machines. As the idea gained relevance in the early 1900s, social workers, government agents, businesspeople, and journalists, among many others, began to use it to regulate everything from fuel consumption to marriage and leisure activities. This machine-based ideology became central to the reconfiguration of the animal body and also informed the efforts of social workers to eradicate "degenerate" behaviors in humans. In the idea of efficiency, agricultural reformers found a solution for scarcity in an age of soil exhaustion, the increase in American urbanites at the expense of the rural population, and the reduction in the number of animals available to fulfill consumers' growing desire for protein. Reformers reconceived animals in terms of efficiency as part of a rational, economy-driven reorganization of the production, distribution, and consumption of American products. Efficiency was not a seemingly benign industrial idea but a loaded concept; it amounted to biotic control over the lives, experiences, and bodies of American livestock.

The notion of efficiency had a "technological heritage" that gave it what scholar Jennifer Karns Alexander calls "objective plausibility." However, efficiency was also bound up with eugenics.[11] The modern, industrial development of American agriculture paired the seemingly objective idea of efficiency with biological hierarchies based on race and breeds, which situated Anglo humans and animals at the top. Reformers did not believe that there was a conflict between these paired ideas. Scientists, economists, and animal husbandry specialists connected the "superior" value assigned to British stock to the "elite" ancestral genetics and also the homogeneity of those animals. Reformers with this efficiency-driven bias for British animals touted these animals as the solution to the problem of the variability of so-called scrub livestock. The use of British livestock to achieve standardization and specialization relieved anxieties reformers harbored about the biological degeneration of America's animals.

In enforcing this paradoxical notion of efficiency, the International played a pivotal role in the remaking of livestock genotypes and phenotypes. The meatpackers used the International as leverage over livestock production, creating an unbalanced or fundamentally unequal feedback loop between Chicago and America's farms and ranches, which aided in the vertical consolidation of the turn-of-the-century meatpacking companies. This provided meatpackers with the unique opportunity to dictate farmer behavior, farm structure, and the types of animals raised in the American countryside. Influenced by the vagaries of animal supply and packer supply-side vulnerabilities, these Chicago businessmen wielded the International to enforce their preferences and shape animals.

Working with professors, who agreed with them as to animal form but differed on the goals for American agriculture, the packers used the International to shape agriculture around an industrial design, opting to prioritize profits over farm or animal well-being and thereby positioning animals as an essential cog in the industrial food production sequence. As a form of technology, the animal was no more or less a machine for converting Corn Belt grains, which were largely unusable as a food source in the harvested form, to human food.

To make machines of these biological beings, professors worked to define what standardization and specialization meant for animals. They inexorably tied single-purpose animals with specialization, deeming dual-purpose animals inefficient on the grounds that they diverted ingested calories to the production of multiple commodities. This diversion of calories prevented them from being the so-called machine that consumed grains of lesser value to create meat of more value once processors extracted it by disassembling and disaggregating the animal. Indeed, packers sold the animal parts at retail prices that were higher than the cost of the inputs. At every stage in this sequence, the row-crop farmer, the cattle feeder, and the packer were required to sell their products in off-the-farm exchanges, making each of them reliant on the animal to convert all these products into surplus revenue. To state it simply, the grain had to leave the farm, the animal could not stay on the farm, and the packer needed consumers, including farmers, to buy these retail products. In this way, the farmer was set on a trajectory to become just as food insecure as nonfarm consumers, and they all were dependent on the animal doing this work for each of them to generate profits.

The notion of standardization accompanied this expected specialization in the commoditization of animality, farmers and packers depended on the consistent replication of food products by the animal. From the valuation of the animal on the farm to consistency in retail cuts of meat, this system relied on homogenous physical forms to extract as much revenue as possible from animal work and bodies. The galvanizing ideology that helped define animal superiority and animal inferiority was eugenics. Like the professors who were concerned about demographic shifts and anxious to improve country life, rural reformers, including President Theodore Roosevelt, worried about "race suicide," which referred to the alleged degradation of the white "superior" races through immigration and the "hyperbreeding" of so-called inferior races, and also about people who possessed defects, including "feeblemindedness," reproducing.

Eugenicists pushed for human sterilization laws, including forced sterilization, marriage laws, and strict immigration laws. With secretary of agriculture James Wilson as president, the American Breeders' Association organized to study and disseminate eugenicism. Racialized belief in white superiority informed the recommendations of eugenicists, many of whom were land-grant professors working at the International, which was reflected in the preference among the organizers of the International for purebred, British livestock with recorded ancestries. The founders of the International believed that these animals were inherently superior to scrub livestock, and their promulgation drove the strong demand for these animals in Australia, South Africa, and North and South America.[12]

Eugenics embodied and emboldened myths and misconceptions that updated manifest destiny under the guise of science, helping rationalize settler colonialism in the West, through which Native American land was expropriated and placed in the hands of white farmers and ranchers. Eugenicists also reconfigured the types of animals that were raised, pushing a version of animal husbandry that characterized animals adapted to the topographies and climates of the South and West as "mongrels."[13] These so-called mongrel animals fit into a broader biopolitical taxonomy that justified seemingly unending harms and restrictions inflicted on ethnic Mexicans, Black Americans, and Native Americans. Racialized and prejudicial tropes that castigated people of Mexican heritage, for example, were also applied to animals, often designated "chihuahuas," or "Mexican," which organizers of the national livestock improvement movement actively sought to exterminate. These myths posited that Hispanic people as well as the cattle associated with them had a propensity to breed at faster rates and that they brought disease and crime—yes, criminal cattle—with them as they traveled. This eugenicism influenced the imperatives of American private business, the research and teaching at land-grant institutions, and federal agricultural policy, as manifested in the US Department of Agriculture's scrub livestock policies in the 1920s.

The department initiated a better sire campaign that was underwritten by this inferior/superior hierarchy. Echoing eugenicists who argued that "inferior" human races were more criminal, better sire campaigns called for "outlaw" train cars to transport "inferior" bulls. To encourage "better" breeding, mock trials were held for scrub animals in which the court charged them with vagrancy, larceny, and theft. Mock prosecutors rebuked the scrub animals, calling them a drain on society for consuming more resources than they contributed. The court always found the defendant—the scrub animal—guilty, and the punishment was execution, illuminating the extermination impulses of the reformers that were deeply rooted in "master race" theories. Repurposing the idea of conquering the "frontier" and manifest destiny, removal and extermination became the tactics used for the overhaul of American agriculture. The moral failings of this racialized understanding of animal quality were reflected in the well-documented agroimperial reach of Britain, the goal of eugenicists to exterminate "inferior" animals, and forced breeding of animals. British livestock were used to replace existing scrubs, casting animals as nuanced, dynamic creatures that could also be objects of human design. Reformers who headed the national livestock improvement movement worked to exterminate, displace, import, and reconfigure meat-producing livestock.

The International also unveiled the economic failings of eugenics in the agricultural arena. The rational, efficiency-driven economic goal of improving the carrying capacity of American farms was undermined by the irrational, cultural aspects of this eugenic worldview, which resulted in real economic and ecological limitations and loss. The Texas tick saga highlighted the consequences of the turn away from diversity. The tick caused a fever that imperiled "improved" cattle, especially as southern cattle were shipped or driven north to market, but the longhorn-type cattle castigated as scrubs were immune to the fever, demonstrating a biological advantage over British animals. And yet these cattle were blamed for the carrying of this disease that hurt and devalued northern cattle, leading the federal government to create a quarantine line that dissected the United States horizontally. Government officials initiated the mandatory dipping of southern cattle in toxic chemicals during the warm months of the year to eradicate the tick.[14]

Limited diversity in animal stock resulted in bioinsecurity and inadequate environmental adaptability—problems that, at least in part, owed to the "master race" mythologies embraced by turn-of-the-century reformers and breeders. Reformers justified their biological preferences by arguing that because British stock had such an imperial reach, demonstrated by the animals' mere presence on many different continents, it must be inherently better or more versatile. Of course, this argument was fallacious: the presence of British stock in so many distant locales was a result of British imperialistic political and economic projects, not a scientifically sound measurement of the biological superiority of their livestock. The genetic homogeneity of these "improved" animals limited their physiological and immunological capacity. Still, narrowing the biodiversity of the modern animal was a central goal to the movement, a process that accelerated following World War II, especially in the pig and poultry industries. Indeed, with each passing decade, livestock represented a more select set of genetics. By the beginning of the twenty-first century, for example, 87 percent of purebred pigs came from only four different breeds—Hampshire, Landrace, Durocs, and Yorkshires. Because of the thrust to eliminate genetic diversity, even several British breeds vanished from farms altogether, including the Suffolk, Cheshire, and Essex.[15]

★

Animal agriculture underwent monumental changes between 1900 and 1920, moving from a range model to a corn-based scientific husbandry practice that fundamentally transformed supply in the meat industry. Many forces converged during this period to instigate these changes, including deteriorated range conditions and other factors that destabilized production, such as the extreme winter weather in the 1880s, economic depression in the 1890s, and the end of the open range symbolized in academic literature and popular culture by the barbed wire fence.[16] The International exemplified a national reform effort, in which both public institutions and industrial firms participated, that promoted and instituted changes to husbandry that ultimately undermined nineteenth-century practices.[17] While many saw animals as static, "natural" beings, it is clear they were biological beings that were reworked and remade, shaped by turn-of-the-century economic, political, and social institutions.

Making Machines of Animals modifies the producer-consumer take on agricultural history by highlighting the motives of the reformers in the national livestock improvement movement who were situated between producers and consumers and by centering the animals themselves, who served as conduits between and among all these actors. Of course, producer needs and consumer demands figured into improvement goals, but the story of the International Livestock Exposition is not complete without an account of the active group of reformers who created the hub in Chicago around which progressive husbandry revolved. For this reason, I largely focus on the *space between* producers and consumers. The following chapters investigate the International triumvirate—meatpackers, professors, and modern animals—and their contributions to the industrialization of American agriculture.

Chicago meatpackers and land-grant university professors created the International Livestock Exposition with two clear goals: to refigure animal genetics and change animals' physical shape. As an overview, chapter 1 describes the emergence of Chicago as the primary meatpacking center and identifies the motives for the collaboration between these seemingly unlikely partners. The packers wanted to control the supply of animals shipped to them, and the professors wanted farmers to be able to increase their revenue and national food output without exhausting the soil. Despite their differing agendas and contributions to the International, the packers and professors agreed that creating animals with uniform genetics and bodies benefited both of their projects.

To shape animals to meet this industrial prerogative, packers and professors first established the genetic priorities of the livestock improvement movement, which is the subject of chapter 2. Reformers preferred purebred animals, and I explore the reasons why agricultural advocates associated purebred animals not only with uniformity but also superiority and why they cultivated the use of British livestock in particular. Practices in the American West reveal nuances in American eugenics that collapsed boundaries between humans and livestock. From quarantines and routine kerosene baths of Mexicans and Mexican Americans crossing the border, restrictive and denigrating immigration laws and policies manifested in the medicalization and militarization of the Mexican border. Not limited to humans, a trans-species, racialized taxonomy of humans fundamentally contributed to the transformation of animal husbandry.

Racialized assumptions about animal bodies and the belief that eugenics produced efficiency in agricultural production was on display at the International. Hoping to regulate animal bodies and to ensure consistency, progressive breeders identified like animals, including close relatives, and bred them to reduce the statistical likelihood of physiological variations. Purebred associations and registries organizationally and administratively supported this movement by cataloging and tracking ancestries and guaranteeing breed purity. Shortly after the inaugural International, the Union Stockyards built the Purebred Livestock Record Building to house these registries and records. By certifying animal "purity," breed associations provided the necessary institutional backing for the International to conduct the show and also to encourage the use of purebred animals throughout the United States, a call that the Department of Agriculture answered in the 1920s with its "breeding up" campaign.

Ultimately, the show ring shaped the tastes and preferences of breeders. Chapter 3 examines how the triad of efficiency, standardization, and specialization manifested in the transformation of animals' bodies, which ultimately led to the reorientation of animality itself. In this case, the judges became the arbiter of modern animal design. The International prioritized a specific set of physiological traits for cattle, sheep, and pigs in its efforts to create machines of animals. These traits linked form to function. More simply, the judges identified and rated body width, depth, and formation, and they associated those traits with a single purpose—meat production.

Land-grant universities played a critical role in the International's organization, but the schools' animals, students, and professors also participated. Chapter 4 outlines these various activities and examines the developing link between the research and teachings of land-grant universities and agribusiness. The most prestigious student honor at the International was awarded to the winners of the Collegiate Livestock Judging Contest. To win the judging contest, students evaluated and placed animals based on the International's standards. Through education, advocacy, and participation, the land-grant partnership helped packers socialize a generation of students and young farmers in the preferences of the Chicago market. The impact of the exposition also radiated out to campuses and influenced student life. The universities held small-scale expositions to prepare livestock and students for the big show in Chicago. University livestock filled the rings at the International and competed against progressive farmers' animals. These show animals required a specific grain diet based on age and production goals. To assist in the transformation of feeding practices, land-grant universities disseminated ideas pushed by packers and professors through demonstrations. In particular, they used educational displays and organized grain competitions to encourage farmers to adopt Corn Belt husbandry.[18]

Finally, chapter 5 evaluates the extent to which the International achieved the goals of the meatpackers and professors who organized it. The show captured the essence of the reform movement and was successful in altering breeders' preferences and animals' bodies. However, even though the show pushed farming toward specialization, it failed to establish a consistent standard in the show ring. To be sure, breeders adopted purebred, British stock and animals' bodies became smaller. But by prioritizing superlative or the "best" animals, the exposition drove producers to extremes. Consequently, the show ring unintentionally encouraged fads in animal type. The livestock market splintered and commercial animals on the farm were bred in accordance with notions of moderation and efficiency while show livestock continued to oscillate between extremes. This gulf between commercial goals and show-animal extremes manifested in the formation of a secondary market for elite, "well-bred" animals.

Nevertheless, the changes in animals and farmer behavior exemplified the centrality and influence of the exposition. The show helped redirect animals toward single-purpose functions, which led to further specialization at other levels of production. For example, the dislocation of organic material from the farm to the market prevented the producers from retaining the requisite biological materials to replenish soil fertility. Scientific management of animal feeds and animal health encouraged even more specialization on the farm. From monocultural crop production to high-density, single-specie protein production, the sort of specialization promoted at the International ultimately undermined the realization of balanced farming advocated by the land-grant university professors.

Livestock efficiency was an important component in the industrialization of the American economy, making animals quotidian biotechnological and mechanized beings. Packers and professors worked with government officials in a top-down way to indelibly link standardization and specialization and to establish the types of animals, the type of farming, and the type of food distribution that increasingly divorced not just urban consumers but also farmers from animality. Following the inauguration of the International, farmers' lives and routines were increasingly devoted to monocultural crop and livestock production, and farmers became more and more dependent on off-the-farm exchanges, decreasing the chances of any individual Corn Belt farmer producing an actual food product. Instead, they produced a biological component that down the line resulted in human food.

This distance between the farmer and food accentuated what the industrial turn meant for animals, their caretakers, and inevitably, the consumer. As the century unfolded, all three of these groups lived and operated in increasingly homogenous settings, interacting with the other only in as much as capital transfer was necessary. Farmers were less likely to own multiple species of animals, and they became more food insecure and dependent on capital transfers and the exaction or extraction of profits from the animal body. This was as true for row-crop farmers as it was for livestock producers; grain farmers were simply situated at a different location in this sequence than producers.

The spectacle that visitors enjoyed at the International highlighted this shift in human-animal relationships—a shift that made livestock shows seem like a zoo where visitors, positioned as onlookers, compensated for their increasing distance from animal life in their daily lives. The International also demonstrated that this shift was not accidental. The national livestock improvement at the International created the conditions for and also benefited from this human-animal void, a direct manifestation of the standardization and specialization of domesticates.

The crowning of champion livestock entertained spectators, while the incentives for producing livestock that accorded with the International's preferences and the penalties for disregarding them resulted in the manufacture of animal bodies that were smaller and more compact and that matured early. These standardized, specialized animals not only required specific farm structures—high-cost inputs, high-density feeding, and surplus, off-the-farm sales—but also forced farmers to rely on a large network of implement and input retailers and nonfarm agricultural experts. These collateral agencies and industries provided secondary opportunities for the meatpackers and agribusinesses to dictate farming practices through agricultural expertise and to generate revenue by selling inputs to the farmer, including fertilizer products, like anhydrous ammonia, that became necessary owing to the removal of the animal from the farm.

The International changed the meaning of animality itself, creating a new biopolitical order connecting the animal to the industrial. Several stages of violence and coercion from breeding to disassembly were implicated in twentieth-century meat production, but animals too dictated human behavior, complicating the simplistic humans-have-dominion-over-animals under-standing of industrial agriculture.[19] The human-animal relationship, of course, was fundamentally unbalanced. But the modern animal shaped farmer choices, dependencies, and fragilities, compelling farmers to dramatically change their routines to accommodate homogenous, high-density animal farming reliant on Corn Belt grains. In this way, both the reformers at the International and the animals themselves helped usher a new age in American agriculture.

Meatpackers and Professors
Take Aim at "Scrubs"

Organization, concentration and concerted action in the live stock
industry will result in wider influence, greater improvement and
prosperity for all interests involved.

J. A. Spoor, "Tells of Great Year"

A towering, central figure in the improved livestock movement, Alvin H. Sanders served as founder, vice president, and eventually president of the International Livestock Exposition. As the editor of the *Breeder's Gazette*, he was
the "chief propagandist" for the movement after 1882, when he took over the
journal from his father, until his death in 1948. Six years before his death, he
published *The Story of the International Live Stock Exposition*, the only comprehensive treatment of the exposition, which details the show's purpose and
impact and describes its varied facilities, animals, and goals. Sanders also
worked assiduously to gain the trust of American political officials. In 1900,
he was selected to represent the United States at the Paris Exposition, and in
1905, he served as the chairman of the American Reciprocal Tariff League,
charged with promoting legislation to expand foreign trade. He regularly corresponded with presidents Theodore Roosevelt and William Howard Taft
on issues both personal and political.[1]

In his various roles as publisher, organizer, and unofficial historian, Sanders developed the mechanisms for the transformation of American livestock.
In *A History of Hereford Cattle*, he describes the nineteenth-century forays
into breeding better cattle and includes an 1883 poem titled "Texas Jane" written by W. E. Campbell, a farmer and livestock showman, that highlights the
qualities that made Texas Jane a celebrity heifer. The poem is written from the
perspective of the heifer, who explains, "My father was a thoroughbred; / My
mother a wild scrub; / The cross makes me easily fed." The purebred Hereford
genetics she inherited from her father put "meat on [her] back," which indicated to progressive breeders and judges that she had a superior carcass. Her
father also bequeathed her aesthetic traits—specific color markings that correlated to superior breeding. In the case of the Hereford, whether crossed or
purebred, offspring were adorned by a white face. The improved livestock

industry and Sanders's reform efforts would mirror the story of Texas Jane, promoting her commercial qualities and aesthetic ideals.[2]

Sanders repeatedly stressed the importance of eliminating poorly bred animals by importing and reproducing British stock throughout the United States. Having written multiple volumes recording the ancestries and benefits of cattle breeds, including Herefords, Aberdeen-Angus, and Shorthorns, he was the foremost American authority on British genetics and the history of improved breeding practices, affectionately nicknamed the "psalmist of husbandry" by his contemporaries.[3] Sanders loathed scrubs and made it his quest to populate the country with more Texas Janes by encouraging farmers to use purebred males to breed better animals, feed animals grain to increase their performance, and select for livestock with meat on their backs.

At the time, agriculturalists used the word "scrub" as a catchall, pejorative term that referred to low-quality carcasses and unknown or inferior genetic makeup. "Eradicat[ing] the scrub," for Sanders, was essential for agricultural advancement, a step that had been taken by purebred Shorthorns, who were the pioneers that conquered the scrubs throughout the United States and exterminated the Longhorn from the plains and mountains.[4] From words like "pioneers" and "conquering" to "extermination," "eradication," and "elimination," provocative language reverberated throughout this movement, indicating an effort to reshape American livestock production by regenerating, through killing and breeding, the animal body.

Sanders unified the meatpackers and professors around one central goal— the elimination of underperforming livestock. Reformers bundled notions of animal value influenced by eugenics, the drive for profits, and the need for food and soil fertility to reimagine the animal. From the late nineteenth century onward, range cattle with their inefficient bodies and "mongrelized" genetics became emblematic of a bygone era. Reformers attempted to justify the changes made to animals by referring to a long list of cliched economics terms—words like "efficiency"—but the so-called scrubs or mongrels portrayed by them as backward, degrading, and even criminal fulfilled important ecological roles through their variability and adaptability, and the eugenic bias for certain breeds or "races" of animals seen as superior often subverted the seemingly rational goals of the livestock improvement movement.

Nineteenth-century range cattle were often some mix of Longhorn cattle, genetic ancestors of the Iberian Longhorn. They mostly came from the Spanish Criollo, populated Florida and Mexico in 1521, and advanced into Texas from Mexico in the eighteenth and early nineteenth centuries. Tough and highly mobile, Longhorns suited the droving and vast grazing demands of the range. They had large, distinct horns and hard hooves that allowed them to fend off predators. They also possessed an immunity to Texas tick fever, and their bodies, including long legs useful for long-distance travel, allowed them to thrive in the subtropical, open-range areas of the United States. These cattle also had high rates of natural increase. They heavily populated the South, the West, and most notably, Texas. Their resilient nature helped them flourish with minimal labor or intervention by ranchers.[5]

Range animals' primary food source came from the roughages available on the land. Producers rarely augmented the animals' diets by feeding them grains, and they depended on unmanaged and seemingly natural pastures. Cattle indiscriminately made use of the range and sometimes wandered as far as fifty miles in any direction. Self-maintenance, nevertheless, came with consequences for the range. Not only did land expansion opportunities for would-be ranchers wane at the end of the nineteenth century, but this herding model also resulted in ecological damage and habitat modification. Overstocking led to soil compaction, selective foraging, and overgrazing near water and salt reserves. Overgrazing led to the killing off of perennials and the destruction of many native plants and fostered ecological problems that affected moisture retention, water quality, and erosion and that contributed to soil exhaustion. In some cases, open-range cattle stimulated desertification. These ecological consequences reduced the carrying capacity, or the number of animals the land could support, over large swaths of the West, which resulted in herd reproductive problems and food output limitations.[6]

Ranch owners invested little in the permanent structures, like barns, silos, or feedlots, that dotted the Corn Belt. Because of limited contact with humans, the animals were often wild and untamed. A "Texas steer" or a "rangy steer" carried negative associations on the market; Chicago buyers ridiculed them. Their brands and behavior reflected the husbandry system in which ranchers raised them. These semiwild animals were given few feed supplements and reached market weights at slower rates. At that point, they often were old, tall, and thin.[7]

By 1900, Chicago meatpackers controlled nearly all aspects of the industry, but they remained vulnerable to supply fluctuations and variations in animal quality and committed significant resources to ridding farms of scrubs. Armour and Company, one of Chicago's major meatpacking companies, for example, created its own research institute to evaluate and disseminate information to farmers. In *Armour's Handbook of Agriculture,* director R. J. H. De

Figure 1.1. An illustration of a scrub bull. *Source:* De Loach, *Armour's Handbook of Agriculture.*

Loach rails against inferior livestock and documents the negative characteristics of inferior cattle. The scrub steer (fig. 1.1) carried excess hide with obvious protruding bones, especially on his top and at the hip, which correlated to poor-performing, low-yielding beef animals. The color pattern also suggested that the animal did not belong to a specific breed. Indeed, the scrub steer lacked recognizable features of any known beef-producing breeds. The Chicago packers additionally associated the larger head shape of older animals with inferior value. The steer's horns consumed valuable space on railcars and injured animals and their meat as well as human handlers. The packers linked these horns as well as their spotted color pattern and weak shoulders and backs with the scrubs of the declining western range.[8]

Disdain for these livestock types linked Chicago meatpackers with land-grant university researchers in the improved livestock movement. To be sure, the packers sought to expand productivity and profits, while the professors worked to improve farm efficiency to increase food output and farmer revenue. Like the packers, Edmund L. Worthen from Cornell University—a soil technology and farm fertility expert—connected improvement to minimizing the use and reproduction of razorback hogs and the "raw-boned, long horned Texas steer."[9] According to Worthen and his colleagues, these semi-wild animals embodied premodern agriculture in their lack of adequate muscle development across their backs, hips, and rear legs—areas where the highest-quality meat was located.[10]

Concerns about animal variability forged a partnership between the packers and professors and masked these reformers' different motives that had the potential to produce conflicts. Power-hungry meatpackers yearned for the ability to directly influence farmer behavior and animal products. Their motives were control- and profit-driven. The professors had broader and seemingly more altruistic goals. These scientists and economists hoped to correct for the shifts in domestic population and meet the food needs of new consumers as well as ensure the viability of American farm life through revenue enhancement and soil improvement.

While the focus on farmers, ranchers, and consumers in the rhetoric of these reformers veiled many ongoing societal frictions, the idea of creating the modern animal triggered few philosophical questions about the merits of transforming animal bodies. This biopolitical imperative transcended meatpacker motives, professor goals, consumer desire, and producer anger and, in fact, was one of the few things that tied these interests together in this era of discontent.

Farmers and ranchers accused meatpackers of manipulating price and overconsolidation, and masses of agriculturalists also greatly distrusted "science" and those who were selling it. Farmers ridiculed and often rejected the suggestions of these "scientific fellers."[11] Consumers for their part distrusted dressed meats shipped across the country and worried about the safety of food. The most powerful packers, the big four, unflinchingly attacked these consumer critics and skeptics in emerging urban markets that they also hoped to woo. But the increase in demand for proteins mitigated consumer criticism, and by the turn of the century, the urban buyer had given way to the Chicago meatpackers. Meatpackers industrialized the disassembly and distribution of meat protein, and consumers eventually embraced the convenience.[12]

Intersecting at the modern animal, these different voices, even when at odds, created the motivation for the national livestock improvement movement, which filled the space between producers and consumers. The apex of this movement was the International Livestock Exposition, which the packers and professors created and wielded to address the persisting problem of animal supply. This beef supply problem revealed an enduring paradox for both the packers and professors. To industrialize meat processing, the packers had worked to protect production from the proclivities of nature, conquering both geographical space and seasonal weather. The professors had hoped to unleash the biological capacities of land and animal to feed an ever-growing nonfarm, urban population. However, to achieve and normalize these industrial ideologies and make them a working reality, the packers and professors had to address the vagaries of animal life. They needed to remake these biological beings in the industrial image, to develop types of livestock that were not affected by the variation, uncertainty, or regional-dependent ecologies intrinsic to animality.

Remaking animal life required a change in farmer husbandry practices, which linked the reformers' macro goals to the micro level—the knowledge, practices, and routines of animal breeders and caretakers. The International became the hub around which progressive husbandry organizations and farmers revolved, an unparalleled force in the national livestock improvement movement. Movement reformers created the standards for the modern animal, founded the organizational means to alter the livestock industry, and established the incentives and punishments necessary to change farmer behavior. The packers and professors normalized industrial ideologies at the International, and the farmers adopted the associated technologies, including the modern animal, unceremoniously. Although consumers and producers were ever present and influenced the motives of the packers and professors, they were largely absent in the formulation of the exposition and the standards created for the modern animal form.

Occupying a position between the farm and the table, the International provided a space for third-party advocacy in a rudimentary feedback loop that dictated and was informed by the growing demands for red meat made by urban consumers and by the political backlash against meatpackers and the economic instability of American farmers and their capital dependency. The packers and professors wielded the International to force changes in ranching and farming to meet the consumerist tilt of American life and redirect the producer mindset toward industrialization. Consensus on the demerits of scrub animals and the need to exterminate them minimized differences among reformers and galvanized their efforts in Chicago. This "amalgamated exposition" in Chicago united the forces of the livestock improvement movement into a singular body.[13]

The Rise of Chicago in Meatpacking

Prior to the Civil War, Cincinnati held the distinction as the primary meatpacking city west of the Appalachian Mountains. Close to the expanding agricultural production of the Midwest and with a developing banking industry and access to rivers that ran both to the East and South, Cincinnati served as a point of connection between farmers and consumers. Because it was surrounded by fertile soil and a growing livestock industry, Cincinnati also attracted drovers trying to market livestock. Cincinnati meatpackers were able to move salt-preserved pork products along the Ohio River on flatboats and steamboats via a system of canals that had been built in the 1830s.

Butchering 150,000 pigs per year in the 1830s and 400,000 by midcentury, the city earned the well-deserved nicknames of Porkopolis and Hogopolis.[14] Travel impediments and minimal upgrades to infrastructure, however, limited the city's advancement in the industry. Most of the live animals moved along roads driven by herders to the market, which was difficult and inefficient. These semiwild hogs with poor herding instincts caused drovers problems. Herders caught the most rambunctious pigs and stitched their eyes shut in the hope of preventing them from running away from the group. Long journeys to market also caused damage to the animal; because of their short stride, hogs lost weight and value during the trip. Cattle and sheep more ably made the drives, but the trips reduced the finished weight of all the animals and bruised or toughened the product.[15]

During the American Civil War, technological innovation and shifting demand pushed the livestock center of the United States west to Chicago, eroding Cincinnati's dominance and allowing Chicago to emerge as a meatpacking powerhouse.[16] The prewar facilities in Chicago were not sufficient, and so John B. Sherman, who had started several stockyards in Chicago prior to the war, partnered with some of his competitors to create the Union Stock Yard and Transit Company in 1865, incorporated with $10 million. They bought 120 acres of swamp ground and expanded to 340 acres by 1896. The owners commissioned engineer Octave Chanute to build a stockyard that could handle all the livestock arriving by train. Chanute remade the landscape south of the city to accommodate the rail traffic, the masses of workers, the hordes of animals, and the waste. He drained low-lying lands and marshes and managed a

Figure 1.2. Union Stockyards in Chicago, 1906. *Source:* MS 506, box 7, folder 4, ILER.

gentlemen's agreement to ensure the cooperation of different railroad and livestock companies.[17]

Many of the primary railroads intersected in or connected to Chicago by the end of the century.[18] Every railroad in Chicago was linked with the Union Stockyards track system. Chanute equipped the stockyards with a web of lines—main lines, side lines, and storage tracks with platforms big enough for workers to unload whole trains at once—that increased the efficiency and speed of the movement of live and butchered animals. In the end, Chanute laid thirty miles of sewers and drains, created a grid of streets that expedited work-related traffic, and arranged five hundred pens complete with chutes, gates, and ramps to move and house the livestock (fig. 1.2).[19] To accommodate the influx of stock, he also created an extensive labyrinth of watering and feeding structures and laid six miles of water pipes. The facility then kept expanding as sales and demand increased for the products manufactured in Chicago. To water the livestock, the owners built twenty-five miles of water troughs, and eventually the facility ran fifty thousand miles of electric wire and had ten thousand incandescent lamps and massive engines for lighting and powering the plant.[20]

The meatpacking industry in Chicago became a technological and commercial marvel, leading observers to tout the Union Stockyards as "the eighth wonder of the world."[21] The owners sought to make the Union Stockyards itself a spectacle, which later allowed city officials and meatpackers to advertise it as a tourist destination. Guidebooks and stockyard manuals noted its size and scale along with its engineering feats. These publications also described its modern amenities, which included administrative buildings, commercial buildings, and a hotel. The hotel, originally named Hough House, featured a 130-foot frontage, six stories, two wings with porches, wide verandas, a billiard room, a parlor, and a barbershop. The proximity of Hough House to Chicago prompted many observers to speculate that the stockyards might well become a retreat for suburbanites and tourists—a notion challenged by other popular images of a meatpacking facility as a place of industrialized slaughter.[22]

Despite these architectural achievements, the Chicago stockyards were a disquieting experience for many spectators. Even though they often witnessed the slaughtering of other species without acute distress, the slaughtering of sheep in particular, during which workers shackled the back legs of the lambs, hoisted them off the ground, and cut their throats to begin the process, provoked sharp emotional discomfort; some even fainted at the sight and were compelled to leave the room.[23] Along with the emotional tumult of animal slaughter, waste and pollution accompanied the advancements meatpackers made in the disassembly of livestock. The packing plants emitted odorous smells into the air and expelled toxic pollutants, stoking serious backlash and a local political movement to control their excesses.[24]

Notwithstanding the consequences of industrial slaughter, Chicago meatpackers remained focused on technological innovation and the vertical consolidation of the industry. The Union Stockyards facilities were both a market for meatpackers to purchase live animals and a place to kill and process them. This consolidation and the capacity to deliver products around the world allowed Chicago meatpackers to market products directly to consumers and thereby undercut competitors, including those in Cincinnati. The development of the refrigerated car for the shipment of dressed meat spurred the growth and dominance of the major meatpackers in Chicago, the meat trust.

Swift and Company, founded by Gustavus Swift, became one of the largest meatpacking companies in the United States. Swift transformed meatpacking technologies, workers' routines, and animal experiences. Dedicated to industrial efficiency, he worried about the risks associated with the transportation of animal meat and by-products. For Swift, shipping animals alive was inefficient. On average, every steer shipped to the eastern markets contained 60 percent waste—only 40 percent of a live steer's body was edible. Swift worried not on about shipping costs but about the risk of damage to or death of animals in the shipping process. Dressed beef—the carcass of the animal after workers kill the animal, remove organs and inedible parts, and hang the remaining intact meat products—traveled more efficiently, taking up less space per unit and allowing meatpackers to squeeze more profit from each animal and train car.[25]

Other companies experimented with refrigerated cars, but their prototypes contained many flaws. George H. Hammond, another meatpacking magnate, sent beef to Boston in poorly designed cars that amounted to large iceboxes on wheels. The packers stored meat dangerously close to the ice, and the dressed meat often came in contact with the ice during transportation when the train turned corners, which ruined or discolored the meat. Also, as the hanging quarters moved, they shifted the weight in the car and sometimes caused derailment. When the trains arrived at stations along the way, workers opened the doors to refill the ice, which triggered dramatic shifts in temperature, potentially harming and spoiling the hanging meat. Meanwhile, Swift developed his own chilled cars by shipping dressed meat in the middle of winter in boxcars, and he instructed workers to leave the doors cracked open to cool the meat and circulate air.[26]

In 1878, Swift hired Andrew S. Chase, a Boston engineer, to update the chilling cars to a standardized, advanced refrigerated car—an improvement that Swift thought would revolutionize the industry and make him fortunes. Chase designed an insulated car with ventilated compartments accessed from the outside for ice and salt. The forced air chilled the meat, and there was no danger of the meat touching the ice or need to expose it to refill the ice compartments. Owing to this technological advance, Swift was able to deliver products to customers year-round and to maintain slaughter operations throughout the summer months. Before the Civil War, July receipts paled in comparison to wintertime, often as little as a tenth of December receipts, because without cooling rooms, meatpackers were forced to slow production in the summer months. By 1880, July packing reached levels of over half of December production, which reflected the vast growth in ice packing and the correlated advantages of maintaining regular output throughout the year, and by the late nineteenth century, weather and transportation no longer dictated the butchering and processing of animals; it became a year-round, industrialized business.[27]

To undercut local butchers, meatpackers exploited the abundant supply of cheap labor and made significant investments in refrigeration. On Swift's heels, Armour, Hammond, and Nelson Morris transformed their methods of butchering to compete, all making investments in refrigeration. The Union Stockyards not only bought and sold animals for all the packers, but Swift, Armour, Hammond, and Morris also centralized production operations. In 1871, meatpackers butchered less than 4 percent of cattle that went through Chicago; by 1883–84, dressed beef surpassed live shipments of cattle for the first time.

Despite the technological marvels created and built by the packers to address the variability of nature, the proclivities, needs, and uncertainties of animal life left them with lingering biological limitations. Fluctuations in the quantity and consistency of supply, for example, frustrated the decades-long attempts of meatpackers to control every aspect of production. Left vulnerable to farmer behaviors and routines as well as the biology of animal bodies, they dreamed of remaking animals and the farm so that they would conform to the industrial ideologies that underpinned the meat trust's business operations.

The Meatpackers

The meatpackers' inability to influence the number of animals sent to market, when they were sent, or the quality and uniformity of those market animals became their chief motivation for organizing the International. The inadequate supply of uniform meat products placed limitations on three objectives. First, the packers sought to expand sales in the urban centers of the United States and foreign markets in Europe and South America. Second, they needed to improve the public image of the meat trust by challenging price-fixing accusations. And third, they wanted to address consumer uncertainty regarding food quality, as that was a prerequisite to expanding their market share. Chicago meatpackers' internal records indicate that these three goals were geared toward one central aim: resolving the meat supply problem.

In the 1895 annual report to the shareholders of the Chicago Junction Railways and Union Stock Yards Company, the board of directors bemoaned the reduction in the number of cattle being shipped to Chicago. Concerned by the increase in the demand for beef that accompanied the growth in urban population, the Union Stockyards rushed to find the cause for this supply problem. After three years of agricultural depression, stock raisers shipped more cattle to the market to make up for lost revenue, which caused a temporary oversupply of cattle on the market and price volatility. Fluctuations in the availability of grain accompanied decreasing farmer revenue. The depression of 1893 and crop failures in 1894 affected corn price and corn supply, placing an additional burden on livestock producers. To supplement income, producers sent production females or brood cows to the market along with steers, which undermined total calves available for slaughter and reduced herd size in subsequent years.

The restocking of herds took years; breeders had to retain female calves to rebuild their herds, which placed an additional encumbrance on the market by eliminating potential finished calves to send to Chicago.[28] As an Armour and Company publication demonstrated, the number of beef-producing cattle shrank as the human population grew (fig 1.3).

During this period, the stockyards executives closely monitored the dramatic decrease in all livestock in the United States. The Union Stockyards Board of Directors reported in 1900 that the number of cattle had decreased by eleven million since 1889, the number of pigs by fourteen million since 1890, and the number of sheep by eleven million since 1883.[29] This supply problem overlapped with an increasing urgency among the Chicago meatpackers to expand production and increase domestic and foreign market share as the consumer class grew and the demand for quality meat swelled.[30] The meatpackers sought to penetrate, control, and command these markets to satisfy their own capitalistic desire for growth and to justify the expansive development of the Union Stockyards. At the end of the nineteenth century, the Union Transit and Stockyard Company had invested heavily in infrastructure updates by building new pavilions, pens, viaducts, tracks, and railcars.[31]

Owing to the amount of control Chicago meatpackers wielded over the market, they became targets of criticism related to high food costs and price fixing. Volatile markets aroused producer suspicion, and high retail meat prices stirred consumer resentment.[32] Charles Edward Russell's volume titled *The Greatest Trust in the World* especially stoked this popular, antimonopoly sentiment. A journalist and political activist, Russell served as one of the founding board members of the National Association for the Advancement of Colored People and won a Pulitzer Prize in 1927 for a different book. *The*

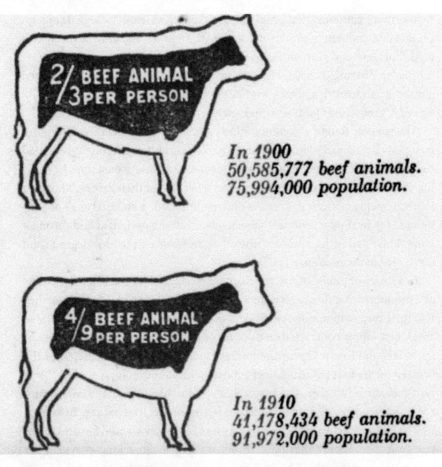

In 1900
50,585,777 beef animals.
75,994,000 population.

In 1910
41,178,434 beef animals.
91,972,000 population.

Figure 1.3. The decrease in beef animals in relation to consumers. *Source:* De Loach, "Beef Cattle."

Greatest Trust in the World originally appeared in serial form in *Everybody's Magazine*. Readers responded with great interest and dismay to the account of the "tragic" nature of modern business relations, which prompted the book-form publication.[33]

No governmental institution, including legislatures or courts, had as much power as the Chicago meatpackers, Russell contends. The meat trust, he argues, owned factories, shops, stockyards, mills, land and land companies, plants, warehouses, politicians, legislators, and congressmen, all of which allowed the Chicago packers to fix prices in a vital industry, one that produced a commodity that every consumer depended on three times a day. Russell furthermore contends that price fixing unfairly distorted the signals of supply and demand sent to producers, warped the market, and unnecessarily created shortages and gluts in animal production. Russell's concerns reflected a broader distrust of packers; journalists, producers, and consumers used similar accusatory language—words like "price fixing," "monopoly," "evil," "greed," "graft," and "loot"—to characterize the power of the beef trust.[34]

The packers fought anticompetition claims and deflected price-fixing accusations by instead blaming inadequate supply and vociferously defended the consolidation of the industry and sidestepped these accusations by shifting the conversation to the ethics, or lack of ethics, of their critics. The meat interest openly criticized journals, newspapers, and "a debased class of politicians" for feeding consumers falsehoods about dressed meat and the meat trust. This "yellow journalism," they despaired, attempted to "degrade and debauch American industries."[35]

In a flurry of publications, the meat interests tried to convince the public and producers that the meatpackers did not influence price by suggesting they had little power to manipulate the market.[36] Time and again articles stated that when supply contracted on the farm or range, the price of buying the raw materials, the livestock, for the packers increased, and then they sold the dressed meats to retail suppliers in consumer markets at higher costs, but at no significant advantage. The "natural cause" emanating from the laws of supply and demand explained fluctuations in the market, the packers insisted.[37]

The Union Stockyards Board of Directors especially wanted farmers to understand the benefits that the consolidation of meatpacking provided the agricultural community. The packers argued that centralization of the meatpacking industry modernized national food distribution networks and thus offered farmers more outlets for commodities, a narrative that attempted to invert the collusion argument made by critics. The propaganda of the stockyards argued that the meat trust was a public good and cast those who distrusted meatpackers as unpatriotic and pernicious; the meatpackers, according to this propaganda, were standard-bearers of modern industrial ideals: uniformity and standardization.[38]

Nevertheless, price and quality drove consumer concern because owing to increased incomes and urbanization, as a 1909 survey indicated, beef had become a larger part of the American diet, and as De Loach of Armour's Bureau of Agricultural Research and Economics observed, consumer demand for quality food accompanied this growing appetite for meat. For consumers, spoiled meat posed a serious health threat, which coincided with a preference for freshly butchered meat. According to Gustavus Swift's son, the idea of eating beef a week or more after workers slaughtered the animal provoked a "nasty-nice horror" among customers.[39]

The outcry from an "embalmed beef" crisis during the Spanish-American War along with the unsavory accounts of the slaughtering and dressing of livestock made popular by Upton Sinclair in *The Jungle* heightened public wariness, which led to federal regulation. Harvey Wiley, chief chemist of the US Department of Agriculture, provided food quality critics with invaluable information on the health effects of preservatives. A champion of government food regulation and considered by many as the "father of the FDA," Wiley conducted a study for the department on human subjects to determine the consequences of using these acids. Wiley claimed that over time, borax and boric acid interfered with digestion and caused damage to the kidneys. The turmoil provoked by Sinclair's portrayal of the meatpacking industry helped Wiley push for regulation and reform in the industry; in the end the department banned these chemicals and approved only salt, sugar, wood smoke, vinegar, pure spices, and saltpeter as preservatives.[40]

Still, notwithstanding the public relations campaign over food quality, the beef shortage was the central problem for the meatpackers, a problem they attributed to inefficiencies in American agriculture. Packers urged farmers to focus on animals' carcass value—uniformity and yield. In doing so, packers' angst about their vulnerabilities to supply manifested in an all-out attack on what they considered to be scrub animals, calling on farmers to turn their attention to "breeding up" livestock.

To address food quality concerns and the supply problem, the packers looked to create an institutional mechanism to normalize industrial ideologies, seeking to change producer business and husbandry practices as well as to remake the American farm animal by pushing farmers to abandon scrub livestock and increase the proportion of improved animals they raised and sold.[41] Enter the International, a "pure food display" that meatpackers hoped would serve as a weapon in this public relations fight, a spectacle that included government inspection demonstrations, dressed meat and meat-product displays, and refrigeration, preservation, and transportation exhibits.[42]

Land-Grant University Professors
What united meatpackers and land-grant university professors was the belief that eliminating scrubs and modernizing the farm was the solution to the supply problem. But the academics' motives differed from the meatpackers'.

Land-grant university officials worried about American demographic shifts and how a seemingly decreasing rural population would be able to produce enough food for a growing urban consumer class. They also worried about the exhaustion of soil fertility as farmers were pushed to produce more food without having access to more land on which to plant. Land-grant professors focused on both short-term national food needs and long-term agricultural production. Soil fertility drove these conversations on demographic shifts, limited acreage, and the modernization of agriculture.[43]

In 1909, Henry P. Armsby, president of the American Society of Animal Nutrition, spoke before a large audience in the Exposition Hall at the International. As the foremost authority in the United States on animal nutrition and experiment station director and department head of the School of Agriculture in State College, Pennsylvania, Armsby represented these professors' concerns about food production and soil exhaustion.[44] The projected growth of the American population, he declared, signaled a potential "deficiency in food supply," and since "new worlds" and the "Old West" no longer offered new land and fertile soil for farmers to till, scientifically informed "permanent agriculture" had to provide for this expanding urban population."[45]

Between 1870 and 1890, the number of farms in America nearly doubled and tilled acreage grew by almost 169 million acres, but this rate had diminished at the turn of the century. At the same time, America's urban population boomed. In the first two decades of the twentieth century, the urban population grew by 80 percent, which represented a vast growth in nonfarm or nonfood-producing consumers.[46] Reformers believed that experiential and technological disparities between urban and rural life drove people to move to the city. By the 1920s, less than a quarter of Americans worked in agriculture, down from over 50 percent in the 1870s. During this same period, farmers made less and less money in comparison to their urban counterparts.[47] Further, producing the nation's food not only required more access to credit and capital but also seemed to many observers to be difficult and unrewarding work. Concerned about rural decline, President Theodore Roosevelt appointed the Country Life Commission to examine agriculture and rural institutions and address, as the commission summarized, "the unequal development of our contemporary civilization."[48] As for the land-grant professors, they tackled this "unequal development" by addressing the push-pull factors of the migration of farm children and young adults to the city, seeking to make life and work in the country seem more attractive. Farm journals highlighted the need to improve farm life and lessen the burden of labor to keep farm children at home. They noted that farm work limited children from using their "wits" and that the development of the brain was left wanting by the extent of the physically exhausting, unskilled labor they were required to perform.[49]

Land-grant institutions addressed the rural problem by utilizing and distributing scientific approaches to farming; they focused on feed rations, the "ideal" physiological traits in cattle, sheep, and swine, and proper land use. Regardless of academic specialization, whether crop production, weed and pest control, animal nutrition, genetics, or meat science, professors argued that it was crucial to educate a persistent class of "soil robbers" by showing them the benefits of modernizing agriculture.[50]

In *The Modern Farmer in His Business Relations*, published in 1899, Edward F. Adams warns readers that ignoring, neglecting, or denying soil exhaustion imperiled national food production. Having worked as a practical farmer, a businessman, and an associate of the University of California school system, Adams feared that farmer prejudice against "book farming" or "scientific fellers" (fig. 1.4) working to transform agriculture into a modern enterprise

Figure 1.4. A cartoon published in 1908 depicting skepticism of the rural reform movement, including scientific or "book" farming, led by President Theodore Roosevelt's Country Life Commission. *Source:* box 21, folder 1, LHBP.

ultimately undermined all farmers and their standard of living. Farmer resentment was understandable; land-grant professors characterized their methods as slovenly, wasteful, and unscientific.[51] However, farmer antipathy toward "book farming" did little to dissuade professors.

The nation could no longer feed itself at the expense of soil fertility—new land did not exist. Land-grant professors maintained that instead of denying science or rejecting it based on prejudice, the modern farmer needed to embrace the challenge of rejuvenating or maintaining "fertility by the operations of chemistry."[52] The seemingly altruistic goal of ensuring high levels of food production for generations galvanized the professors to assume the responsibility of convincing farmers.

Soil scientists T. Lyttleton Lyon and Elmer O. Fippin of Cornell University held that soil management was a solemn responsibility of the university and of the farmer; the "man who owns and tills the soil," they assert in *The Principles of Soil Management*, their evaluation of soil robbing, has "an obligation to his fellowmen for the use that he makes of his land." This was a two-way street, however; nonfarmers had the responsibility of ensuring that the farmer made enough money so that he would not be driven to "rob the earth in order to maintain his life."[53]

The land-grant professors pushed balanced farming—a husbandry regime that required livestock and crop production to coexist on every farm.[54] Under the balanced farming model, modern farmers fed grain to livestock for the surplus production of meat, and then they distributed the livestock waste—primarily manure, used bedding, and wasted feedstuffs—onto fields to improve fertility. While commercial fertilizers, sometimes manufactured and sold by the meatpacking companies, were increasingly available, they failed to provide the requisite organic matter and soil structure to maintain yield goals.[55]

Permanent agriculture was another objective of the professors' reform movement, which they proposed as the replacement for the extensive farming model of the nineteenth-century range that depended on new land as soil fertility diminished.[56] Owing to decreasing access to free grass, free water, and free land, to the ecological consequences and habitat destruction caused by overgrazing, and to what the professors saw as the inferior animals that stocked the range, the output of food per acre or yield per acre fell short of the growing needs of the consumer class.[57]

Agronomist Cyril G. Hopkins, vice president of the experiment station at the University of Illinois, a prominent advocate of permanent farming, and the leading authority on farm fertility, maintained that the broader land-grant and experiment-station community were the "guardians of American soil." According to Hopkins, figuring out how to use land without abusing it was their central objective. No country dependent on extensive agriculture, he contended, ever produced food without exhausting the soil. These practices, he argued, "ruined land," and the science of agriculture would restore it. Multigenerational farm profitability and food output requirements necessitated the belief in and practical application of science on the farm—an intensive farming regime that coupled mixed crop husbandry with modern livestock.[58] Ideally, farmers would then feed these varied crops or crop products to livestock.[59]

Finally, without improved animals, the modern farm would be incomplete. As Armsby declared at the International, "inferior animals," whether fed correctly or not, performed too poorly to advance the cause of scientific agriculture.[60] Improved animals would allow farmers to increase the surplus production of human food, which in turn would improve national productivity per acre. The professors' concerns regarding soil fertility and food output circled back to corralling and systematizing animal life just as the packers' worries over supply did.

The Packers and Professors Meet

Despite their different motives for creating and sustaining the International, meatpacking and land-grant university officials worked to create an exposition unmatched in scale and quality to serve as the central mechanism for improving meat-producing livestock, which for them meant first and foremost eliminating scrub animals.[61]

Following John Sherman's retirement at the end of the nineteenth century, J. A. Spoor became the president of the Union Stock Yard and Transit Company. Spoor believed that problems in the industry could be addressed by the meatpackers, and he confided in Arthur G. Leonard, manager of the stockyards, who energetically crafted intervention plans to promote the production of better stock.[62] He first created a plan to send "missionary bulls" to western ranges. He hoped to create a sort of cooperative for genetics by distributing a well-bred bull to each farming community that could then be shared among breeders. He quickly gave up on this difficult-to-execute venture, but his failed attempt to reform animal husbandry primed him to hear bold ideas. In 1899, Sanders, Robert B. Ogilvie, William E. Skinner, Mortimer Levering, and G. Howard Davison concocted a plan while they were at a livestock show in Toronto for a transformative national show, unmatched in scale, that would be underwritten by the Union Stockyards. Upon return, they approached Leonard with the idea, and he fully endorsed the proposition. This show, they believed, would be the ideal mechanism through which to intervene in animal life and disseminate industrial ideologies.[63]

Leonard provided his unwavering support, which made convincing Spoor an easier task. Sanders remembered that Spoor gave Leonard the "'full steam ahead' signal," and the stockyards quickly sent out invitations to land-grant university officials and faculty, national breed associations, and many in the agricultural press. In November 1899, top agricultural officials flocked to Chicago for a mass meeting at the stockyard's Exchange Building. Levering from the American Shropshire Registry Association called the meeting to order and appointed Skinner as temporary chair of the event. Skinner, who served as a general agent for the Union Stockyards, represented the meatpacking companies' interests along with Spoor and Leonard. Following Skinner's appointment as chair, the body elected Levering as secretary.[64]

Sanders presided over the proceedings, and he began with a vote to authorize the permanent organization of the group, which became the International Livestock Exposition Association, and then he proceeded with the nomination of officers. The interests of the organization—those of the Union Stockyards—were reflected in who was elected officers and directors, which included J. Ogden Armour and E. F. Swift, and Spoor, who took the reins of the association when he was elected president. Dewitt Smith, who resided in Springfield, Illinois, and formerly served as president of the Consolidated Cattle-Growers' Association of the United States, was made first vice president, while Sanders was second vice president. Skinner became the general manager. The board also included land-grant university and breed association representatives.[65] The interests of the breed associations overlapped with those of the professors; land-grant university officials organized and ran many of the breed registries. The most prominent among them was Charles F. Curtiss, who served as an International founder and official livestock judge as well as a breed association organizer and professor of animal husbandry and dean at Iowa State College.[66]

After deciding on the roles and the powers of each office and director, the executive committee concluded that the show would be called the International Live Stock Exposition and would take place in December. Sanders referred to the people at this meeting as the "founders of the International," and he emphasized that their goal was improving livestock.[67] Progressive animal agriculture required the "extermination" of so-called scrub livestock.[68] To ensure quality, the founders created a sift committee whose job would be to eliminate poor livestock before the competition began to guarantee that the International only displayed improved, modern animals.[69]

Although the founders had high hopes for the International, they recognized that transforming animal agriculture by reeducating farmers and altering animals would be difficult. Thus, Sanders referred to the first International as a "trial balloon." By the second year, over four hundred thousand domestic spectators attended, along with visitors and dignitaries from Argentina, Germany, Italy, Belgium, Japan, Mexico, and Central America. American political officials also traveled to Chicago for the festivities, including secretary of agriculture James Wilson and governors from Illinois, Minnesota, Nebraska, Iowa, and Michigan. Sanders declared that "public-spirited citizens," premier farmers, land-grant colleges, and the stockyards had shown a deep concern for "our greatest single national industry." These trial expositions provided enough evidence to make it permanent.[70]

The construction of an enormous new exposition building sealed the founders' commitment to keep the show going. To raise the capital needed to fund the construction plan, the directors and Sanders offered annual or lifetime memberships to the International Livestock Exposition Association. Despite construction delays, which required organizers to postpone the fifth International for two weeks, they finished and unveiled a large hall on Halsted Street in 1905.[71] The amphitheater provided stadium seating for ten thousand spectators in a facility that looked like a capitol building or a sports arena; at the time, no other construction devoted to livestock expositions rivaled it. Built of cement, brick, steel, and glass, it contrasted with the wooden barns on the farm and the horse stalls of the turn-of-the-century city.

It represented a new age in agricultural production and included the modern amenities of the emerging metropolis (fig. 1.5). Steam pipes heated the auditorium and ran by the feet of the spectators, providing a comfortable experience in the cold and windy climate of a Chicago December. Incandescent, regular arc, and blazing arc lights illuminated the arena for the nighttime events that drew maximum capacity crowds. The stalls for the animals shown at the International offered conveniences unknown to many contemporary humans, let alone animals. Their steam-heated, electric-lit quarters provided a stark contrast to those of the slaughter animals on the very same grounds.[72]

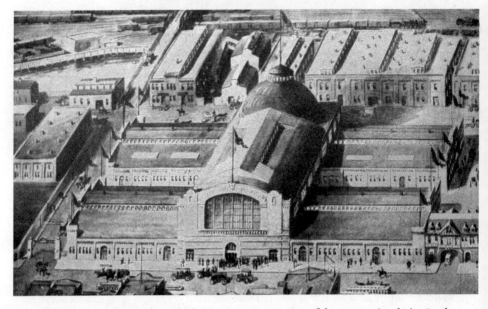

Figure 1.5. International Amphitheatre. *Source: A Review of the International Live Stock Exposition*, 1916.

Consumers and Producers

This biopolitical project masked many of the tensions and conflicts during this era of discontent. Aiming to satisfy consumer lust for meat and encourage farmer adoption of new technology, the packers and professors overlooked philosophical differences to remake the animal form. In the early twentieth century, consumers enjoyed cheap and accessible beef at ever-increasing rates. Even though physically absent, the urban consumer was ubiquitous in the livestock improvement movement. Through geographic and demographic shifts, expressions of outrage and demand for more meat of a higher quality and uniform shape, consumers molded the motives of the meatpackers and the goals of the land-grant professors.

The meatpackers felt empowered to reconfigure animals so that cuts of meat of all beef-producing cattle, mutton-producing sheep, and pork-producing pigs were the same type and size and had the same look in order to satisfy the aesthetic desires of the meat-counter shoppers. The biggest obstacle to this goal was the animal itself. Animal variability based on age, geographical and ecological situation, and breeding as well as farming regime made achieving this consumer-driven objective difficult. The Chicago meatpackers established meatpacker-financed research centers, not so different from government-funded experiment stations, to aid in the project. Armour's Bureau of Agricultural Research and Economics, for example, gauged and shaped consumer preferences and reoriented farmer behavior and animal bodies accordingly.

Livestock producers also participated directly in this feedback loop with the reformers at the International. Discord in the agricultural community escalated from the 1890s through 1920s. Farmers distrusted banks and big companies, including the meatpacking firms in Chicago. With land values and input costs increasing and commodity prices fluctuating, causing intermittent and often severe agricultural depressions, farmers created new political organizations and cooperatives through which they were able to gain access to credit, influence money supply, and improve their bargaining power. Despite their anger and their suspicion of "book farming," many farmers embraced the technologies, including modern animals, associated with "scientific farming."[73]

Even though producer needs and consumer demands influenced the packers and professors, they operated as third parties in their spearheading of a national reform movement to transform the American farm, the meat-producing animal, and the aesthetics of meat cuts on the retail counter. Through the International, the meatpackers would thoroughly incentivize the reproduction of the "ideal" meat-producing animal, while the land-grant professors defined the forms, types, and genetics of the modern animal. The goals and the structure of the International, including class format and judging preferences, would revolve around two biological configurations: purebred genetics and standard body-types.[74]

CHAPTER 2

Breeding Up Livestock

The wide difference in market value between scrub stock and the
improved breeds of live stock caused those interested in the development
of the industry to cast about for some comprehensive means of forcibly
presenting to the farmer and feeder the folly of producing inferior animals,
when the same amount of feed and labor expended on the better kinds
would bring a greater measure of profit.

A Review of the International Live Stock Exposition, 1913

In 1902, on behalf of the Theodore Roosevelt administration, secretary of agriculture James Wilson offered lusty support for the goals of the Chicago
International Livestock Exposition. During the unveiling of the newly built
Purebred Livestock Record Building, Secretary Wilson proclaimed that
the International was "the most magnificent expression of progressive
breeding . . . in the history of any country."[1] The secretary endorsed its objectives and contended that its mission—domestic animals that embodied the
elements of "progress"—contributed to the overall well-being and competitiveness of the nation. Wilson also argued that the International effectively
supported the work of the public land-grant universities and experiment stations. The International, he said, pushed "farmers toward the establishment
of the science of breeding," which required them to raise the type of animals that increased the efficiency of food production and minimized waste.
Indeed, as Wilson both declared and predicted, the International served as
a hub for land-grant universities and experiment stations, along with Chicago meatpackers, in the dissemination of new husbandry practices. "What
a long time we have waited for all of this!" he exclaimed to the crowd.[2]

With the full support and involvement of the US Department of Agriculture, the Union Stockyards advanced the cause of purebred livestock. The
stockyards housed the National Breeding Record Associations of the United
States and also provided breed associations offices and space to store official
pedigrees in the Purebred Livestock Record Building.[3] A year after Wilson
honored the unveiling of the building, Alvin H. Sanders worked with Arthur G. Leonard, manager of the Union Stockyards, and Robert B. Ogilvie,
secretary of the American Clydesdale Association and head of the Interna-

tional Horse Department, to organize the Saddle and Sirloin Club, an exclusive society devoted to modern agriculture.[4]

To compete on a national scale with the prestige of the steel and rail industries, the Saddle and Sirloin Club publicized the International's important reform work. It brought together rural agrarians and young students with Chicago businessmen, national policy makers, foreign dignitaries, and scientists, "serving as a university in the highest order." The organizers believed that "every young man interested in any phase of animal husbandry should become a member." And the members utilized the plush setting, the grandeur of the halls, and the sophisticated architecture to magnify the importance of the society's goals. Baronial Hall, for example, featured a vaulted oak-beamed ceiling and dark-paneled walls reminiscent of banquet halls in medieval England. The club included a series of opulent rooms—an English dining room, a lavish smoking room, and a world-class agricultural library—that exhibited oil portraits of hall of fame inductees.[5]

The most important room in this posh club was the Sanctum Sanctorum (Latin for "holy of holies"). Club members used a wide variety of religious-based words to explain the importance of this "holy" room and the club itself. The room featured "a Pantheon" of portraits of individuals whom the club considered the forebears of the improved livestock industry. Club members touted Britain as the nursery for superior stock, and they hoped to stimulate the use and importation of improved stock, providing American reformers with the genetics needed to redirect livestock production and eliminate the scrub in the United States. Chief among their British "heroes" was breeder Robert Bakewell. Bakewell created an economically competitive sheep that he bred for meat production rather than for wool, milk, or fertilizer, which was the norm at the time. He systemized breeding and selection, marking the beginning of this effort for the International's founders. He was one of the first to be enshrined in the Sanctum Sanctorum, celebrated by club members for his foundational contributions to purebred animals and to breeding practices of succeeding generations.[6] The American Breeders' Association featured him in its very first volume. His influence as an "originator" of scientific breeding and "creative" animal selection made him a legend among members of the association, including Secretary Wilson and Charles F. Curtiss. These reformers appreciated Bakewell's emphasis on increased productivity.[7]

While the club complemented the work of the International by manufacturing a tradition that honored British husbandry and British animals, an aggrandizement that was seen as key to convincing farmers to adopt "superior" biological beings and farming practices, the packers and professors pivoted toward the twentieth century's modern economy, rejecting the pastoral agricultural past, including range husbandry, and working to free agriculture of so-called inferior livestock by remaking farms and animals in the industrial image.

In Chicago, the meatpackers created the first modern production lines. The disassembly of animals predated the industrial titans of the twentieth century, including Henry Ford. But these disassembly lines were limited by animality itself. Animals across the United States lived in different ecological spaces with different bodies, developing at different rates. The industrial turn required reformers to eliminate these irregularities of animality, a biopolitical effort informed by eugenics to achieve specialization, standardization, and efficiency.

Sanders pushed farmers to adopt what he considered genetically superior British animals; because of known parentage, purebred livestock, he believed, produced more standard, predictable, and competitive types of animals.[8] The uniformity of British-based Herefords, for example, resulted from over a century of ancestral "purity"—a familial relationship that established unchanging type and character.[9] Curtiss, professor of animal husbandry and dean at Iowa State College, also linked progressive agriculture to British husbandry. The use of purebred livestock, Curtiss argued, eliminated inefficiencies in food production.[10] The packers shared this conviction—J. A. Spoor, president of the Union Stockyards, argued that purebred animals possessed tangible benefits in genetic predictability that were crucial to achieving the goals of improvement.[11]

Purebred animals imported from Britain addressed both packer and university researcher concerns about agricultural production, a problem they believed could be solved through the "blood" of livestock.[12] Furthermore, through generations of selective breeding and inbreeding, purebreds guaranteed a more standard product. British breeds, whose biological outcomes were predictable and who had specific climatic and nutritional needs, worked well with the industrial goals of standardization and specialization. Spoor and Curtiss also associated this sort of uniformity with superiority, and they referred to it in creating a biological taxonomy that located British livestock at the top and the Longhorn-based cattle of the South and West at the bottom.

The eugenic belief that Europeans from the northern and western parts of the continent were superior to those from the southern and eastern parts underwrote this animal taxonomy. The lingering moral question raised by this racialized taxonomy was not whether or not British purebreds were in fact as uniform in physiological type as promoted but whether they were superior and that other animals, correspondingly, were inferior and deserving of extermination. Nevertheless, the eugenic drive behind livestock reform persisted at and was normalized by the International through its hosting and partnering with a national eugenics organization, the American Breeders' Association, and several nationwide campaigns to disseminate eugenic ideas among American farmers. To institutionalize this uniformity-driven mantra, purebred livestock registries recorded ancestries and worked at the Union Stockyards to aid farmers with genetic selections that reformers believed would eradicate defective or undesirable traits. Consistency in genetic selection that manifested in dependable types of animals benefited farmers by increasing the reliability and value of their products and helped meatpackers satisfy the food quality demands of urban consumers.

The International positioned itself as the mediator between this fervent belief in the superiority of British purebred animals on the one hand and the reality of crossbred or even "inferior" livestock prevalent among commercial operations on the other. While the International wanted farmers to raise animals with purebred parentage, both male and female, of the same breed, practical limitations, especially farmer resistance to abandoning all grade or crossbred livestock, made it difficult to achieve this goal. However, small farmers could still reap the benefits of improved "blood" or genetics by buying or renting just a single purebred sire. Some competitions at the International and "breeding up" programs focused on the use of a single sire to make this effort more affordable and practical for the average farmer. This movement's focus on purebred sires became a national "crusade" when the US Department of Agriculture launched its better-sire campaigns, which featured the public execution of inferior male livestock.[13]

The International served as a gathering ground for those working to wrangle the variability of animality and remake livestock. The International was more than an event, it was also an institution. The International connected the farmers and reformers at the yearly exposition to nationwide movements devoted to eugenics in the United States that consolidated humans, plants, and livestock around similar ideologies about genetic purity and aesthetic and physiological similarity—ideologies that linked cultural assumptions about these biological beings and then sorted them into superior and inferior categories. The International not only provided the mechanism for sorting livestock into these classes denoting biological value, but as the center of reform work, the exposition was instrumental in normalizing these cultural assumptions as well.

The Foundations for "Superior" Genetics

The International became the standard for measuring improvement in progressive breeding; winning came with honors and recognition. The first champion steer in 1900, Advance, became an instant icon (fig. 2.1). Farmers knew of Advance's preeminence, and he became a focus of advertisements for those pushing purebred genetics.[14] The Aberdeen-Angus breed association featured Advance to impress on producers the importance of record keeping and purebred ancestry, as well as the superiority of the Aberdeen-Angus among meat-producing breeds. Purebred breed associations competed with each other to encourage the spread and influence of their breed.[15]

Advance received top reviews from the International judges. Curtiss judged the Aberdeen-Angus steer competition. He commended Advance for demonstrating excellence in both quality and uniformity. Advance carried in

Figure 2.1. Advance, "The World's Champion Steer." *Source: American Aberdeen-Angus Herd-Book* 11, 1901

his blood elite genes from an inbred cattle family—a close relative of Advance won champion at the World's Columbian Exposition at Chicago in 1893, Young Wellington, the son of imported Wellington.[16] Advance merited particular acclaim, Curtiss argued, for having the type of body—stout with adequate flesh—best "for the butcher's block." Advance was first among a large contingent of Aberdeen-Angus that garnered high prices at the International sale. This breed of "market toppers" was featured as ideal beef cattle that converted feed to meat at a high rate and that was easy to flesh.[17]

Not only did Advance gain fame for his elite genetics and success, but his accomplishments also made his owner, Blanford R. Pierce of Illinois, a celebrity. The breeding of select Angus resulted in his nomination and induction into the Saddle and Sirloin Club Hall of Fame. The club commended Pierce for his dedication to purebred livestock. In the club's official biography of Pierce, it celebrates him for importing elite or "superior" bulls from Britain despite the cost, noting that his ambitious determination to obtain the best bulls had a broader impact on the industry. These reformers believed that the generational benefits gained by importing "superior" genetics more than compensated for the high cost incurred by Pierce—Prince Ito, for example, cost Pierce $9,100. The elite bulls, rams, and boars bought for improvement dramatically outstripped average values. At the turn of the century, the average value of an on-the-farm bull over one year was $34.49 and the average value of all livestock on each farm was $536.[18]

With this prestige, Britain became a sort of foundation farm or "stud farm" for the Western world. Britain sold elite, well-bred stock to the United States— the transfer of biological technology in the bodies of animals—and in return, American producers sold commercial livestock and animal products to the island country.[19] In 1900—the same year the esteemed judge and professor awarded Advance top honors at the International—Curtiss traveled to Great Britain to observe livestock expositions and to inquire into the purebred livestock trade. Curtiss began his investigation while on a "cattle boat." The boat carried 657 cattle and 999 sheep for the Liverpool market, and in an article for *Live Stock Journal Almanac*, he assures readers—producers interested in buying livestock from Britain—that the shippers successfully and safely transported livestock across the Atlantic Ocean.[20]

Curtiss details the experiences of the animals to foster confidence among potential American buyers, lauding British producers for providing the Western world with "superior breeds of live stock," and reminding his audience that the success of these farmers in animal husbandry did "not come by chance." Instead, British livestock productivity resulted from "practical object-lessons and scientific illustrations." Unlike many American producers, he observes, the British had "the almost universal interest displayed in matters pertaining to live stock, and the general intelligence characterising this interest." He also applauds the British farmer's use of a balanced husbandry regime that ensured the continual fertility of the soil through the application of manure as a nutritional supplement, thus enabling farms to be "permanently successful."[21]

Late nineteenth-century reformers, including James Sanders, Alvin Sanders's father, used the influence of the *Breeder's Gazette* to get trade barriers on breeding stock removed and thereby guarantee constant access to British livestock. Sanders reviled customs duties; he thought restrictions on trade damaged agriculture and limited the "national wealth" generated by importation.[22] In 1887, he circulated an article about *United States v. One Hundred and Ninety-Six Mares*, in which the court upheld a statute that protected livestock importers from paying a duty on the value of the breeding animals purchased. The statute, intended to encourage the advancement of agriculture, provided this loophole to farmers who bought animals only for "breeding purposes," and the court sanctioned this duty-free arrangement.[23] For agriculturalists attempting to transform the industry in the twentieth century, importation was crucial: advancements made on the farm directly depended on the gains made through the purchasing of elite, British stock.

While access to regular shipments of livestock from Britain helped progressive breeders regularly participate in this transatlantic trade, the extent of sales to American breeders stoked protectionist concerns among the British breeders. Despite the financial gain, British farmers worried that this trade relationship could deplete their genetic advantage over other countries and that American farmers equipped with many of their best animals would undermine British competitiveness. American farmers, many British exporters fretted, would edge out British farmers in the sale of purebred livestock to producers in the United States. This trade relationship was, however, viewed very favorably among British authorities. Sir Richard Powell Cooper, a veterinary surgeon and entrepreneur, argued that British stock had greatly contributed to the improvement of foreign and "colonial" animals. While he conceded that with the establishment of purebred farms in other countries, British farmers' newest competitors would be their former customers, he encouraged breeders, urging them to keep up the work of improving stock, as that would ensure that the demand for their great stock would continue. By maintaining a competitive advantage and advertising in foreign markets, Cooper believed, the demand for British farmers' well-bred animals would remain strong.[24]

Those who aggressively embraced this trade relationship moved to take advantage of it. The success of British livestock in the International Livestock Exposition prompted breeders to advertise and directly appeal to American customers. Inside and outside the United States, the International served as the greatest measure of agricultural advancement in the Western hemisphere, and British breeders used the show's prestige and these transnational successes to feature and sell their animals. The Babraham Southdown flock of Cambridge, England, for example, became one of the most popular flocks after successes in the United States as well as at the International Exhibition at Paris in 1900.[25] To amplify and leverage the success of this flock into sales, its breeders used the International to circulate farm advertisements and boast about their accomplishments around the world. By 1910, they had sold rams and ewes to producers in sixteen different countries, including the United States, Chile, Argentina, Japan, New Zealand, and Russia.[26] The Rowe brothers raised horses in Maple Park, Illinois, and Fishguard, South Wales, and they exhibited their horses in Chicago and London. They advertised extensively at the International, touting their animals to American producers and the benefit of having a horse seller with established farms on both sides of the Atlantic.[27]

The International institutionalized the cultural, or eugenic, and agronomic preference for not just purebred animals—livestock with similar, documented, and replicable ancestries and standards—but specifically for breeds that originated in Britain. With cattle and sheep in particular, the International very narrowly defined the breeds that qualified for competition, limiting producers in the presentation of British breeds only. For this reason, a sheep breed of African origin, Tunis, was excluded from the International. Originally imported at the turn of the nineteenth century, Tunis were raised by many prominent political officials on the Eastern Seaboard, including President Thomas Jefferson. With their low feed requirements, resilience, and prolificacy, Tunis fared well on seagoing vessels, including pirate ships, as well as in the different topographical and ecological regions of the United States, and they were palatable to the American consumer. Even at the turn of the twenty-first century, when Tunis finally started to gain entrance into all the major American livestock shows, judges and breeders ridiculed the traits that made the breed unique and useful and pushed breeders to reshape their bodies to meet the standards set by British breeds. These lingering preferences were historically rooted in a sense of genetic superiority underpinned by "master race" theories that were embraced by agricultural reformers and eugenicists at the American Breeders' Association in the early 1900s.[28]

Curtiss worked with secretary of agriculture James Wilson and Willet Hays, Wilson's top assistant, to provide organizational support for the investigation into the value of genetic selection in both animals and humans. Curtiss, Wilson, and Hays formed and served on the association's board; Secretary Wilson was the president. Many of the land-grant professors who contributed to the founding of the International also participated in the creation of the association and in the promotion of eugenic ideas. The association held annual meetings to address wide-ranging topics on genetic selection, including commercial corn breeding and the importance of Mendel's Laws on animal reproduction, and it also published the *American Breeders Magazine*, which later became the *Journal of Heredity,* with the goal of broadening the discussion regarding selection and to share Darwinian and Mendelian principles with farmers and professional agriculturalists.

Per its devotion to eugenics, the publication urged not only the reproduction of desirable traits but also the elimination of unwanted characteristics.[29] James Poole, a Chicago livestock market expert and journalist, echoes the language of eugenicists in an article published in 1922 reflecting on the early success of the International Livestock Exposition. Like the eugenicists at the association, Poole touts "master race" theories when he derides scrubs as "mongrels" and "Mexican cattle." These types did not possess the beef-making qualities of the International's stock, according to Poole; they lacked muscle shape, docility, and fecundity. In contrast to some of the packers and professors who expected that "breeding up" with good bulls would eventually solve the problem, Poole took a more extreme stance, embracing the extermination and elimination ideology of many eugenicists, arguing that a good bull "cannot atone for the faults of angular, ill-bred Chihuahua cows." His use of the term "Chihuahua" is instructive; this reflexive, pejorative word no doubt struck a chord with many readers by playing on their racialized biases, and Poole's use of it reflected federal and state policies that discriminated against ethnic Mexicans. These types of animals were a major barrier to improvement in the industry, Poole concludes, because of the "infestation" of the range with these "Chihuahua cows."[30]

Poole's sneering and reflexive use of racist language mirrored a broader movement in American politics. Every level of government in the United

States moved to provide statutory support for eugenics. After Reconstruction, the institutionalization of a racial hierarchy was codified in the South with Jim Crow laws. Sterilization became official policy in 1907 in Indiana, which was the first of many states to implement sterilization laws in the United States. The US Supreme Court addressed sterilization in *Buck v. Bell*, a decision in 1927 that upheld the constitutionality of sterilization in Virginia. Eugenics played a part in the Johnson-Reed Immigration Act, a federal law enacted in 1924 that restricted immigration to the United States based on a racialized taxonomy. Through sterilization, immigration control, racial separation, and antimiscegenation marriage restrictions, Americans of Western European descent worked to address what they described as racial decline and degeneration.[31]

The embrace of eugenics in the Southwest, in particular in border towns like El Paso, resulted in the harsh quarantining, chemical disinfecting, and violent treatment of ethnic Mexicans. In fact, these harsh chemical baths, which often resulted in birth defects, cancer, and death, prefigured the violent procedures Nazis used in the Holocaust.[32] Eugenic reformers justified the militarization of the border and these so-called medical procedures through a litany of racialized assumptions about the danger of immigrants that conflated the racialized other with the "enemy other."[33]

David Starr Jordan, who served as president of both Indiana University and Stanford University as well as the first chair of the American Breeders' Association Committee on Eugenics, worked with his colleagues, such as Alexander Graham Bell, on the committee to apply eugenics to humans, plants, and animals. Jordan adhered to a set of racialized beliefs embedded in eugenics that cast ethnic Mexicans as the "enemy," justifying and manifesting in, as scholar Alexandra Minna Stern has argued, "aggressive disinfection rituals . . . based on exaggerated, nearly hysterical, perceptions of [migrants] as dirty and diseased." Not only were Mexican migrants portrayed as a public health risk, according to Stern, but they were also seen as "hyperbreeders" that "threatened to drain public resources."[34] Jordan became one of the most famous eugenic leaders, helping to create a biopolitical order based on a sense of white racial superiority.[35] Jordan's description of "hyperbreeders" echoed Poole's characterization of so-called undesirable livestock. In fact, eugenicists, whether they studied animals or humans, relied on similar racialized taxonomies, as witnessed in the fact that Jordan and Poole used similar language to castigate ethnic Mexicans and so-called Mexican or Chihuahua cattle as inferior and causing infestation.

Westward expansion during the nineteenth century and the racialized violence that cleared the so-called frontier for white settlers directly influenced
beliefs about what animals and what people belonged on the range. "Improved" livestock filled the void left by the killing of animals, including the
buffalo, that were necessary to sustain life for ethnic Mexicans and Native
Americans. These British animals thus aided in settler colonialism, including
one of the most popular breeds of cattle featured at and promoted by the International, Shorthorn, otherwise known as "the universal intruder."[36] Settlers and policy makers used the bodies and genetics of this breed to transform
colonial spaces and husbandry regimes as part of a broader project that connected the biopolitical expansion of the US government and white settlers to
other spheres of the British Empire. The removal and extermination of ethnic
Mexicans and Native Americans and their animals as well as the expropriation of their land morphed into and was "scientifically" justified by eugenic
projects.[37] These racialized ideologies were never separate from the soil, and
they informed the conservation and preservation movements of the early
twentieth century. Eugenicists like Jordan believed that as the so-called
conquering people from the "superior" races of Anglo-Saxons and Nordics
populated the West, they maintained and protected the land and natural
resources, thus securing a chief aim of the professors at the International.[38]

The eugenic denigration of ethnic Mexicans' animals was encapsulated in
the ongoing ridicule of the Longhorn despite the breed's adaptive strengths
in the South and West. Their horn size, long legs, wily demeanor, and hardi-
ness, which the packers and professors claimed made them inferior and inefficient, were suited to the subtropical regions of the United States, allowing
them to thrive with little intervention. Longhorns' lean bodies and long legs,
helped ranchers take advantage of the vast grazing grounds and the unique
ecological qualities of the warmer climates.[39]

In addition, the breed was largely immune to Texas fever. In their southern habitat, Longhorn cattle experienced little damage or illness when they
came into contact with the cattle tick that transmitted the fever, a species of
tick that only existed in areas of the country that did not see a good winter
freeze, which killed it. Northern livestock were vulnerable to the ravages of
the fever. The disease attacked the animals' red blood cells, and farmers often
noticed blood in their cattle's urine. After exposure to southern cattle, infected
northern stock usually died in under ten days, and as much as 90 percent of
an infected herd could succumb.[40]

The concern about the Texas tick threat to northern cattle led to containment of the Longhorn. The United States became divided agriculturally based on immunological responses to Texas fever, when in 1890, the secretary of agriculture quarantined southern cattle. The initial quarantine line covered Arkansas, Texas, and the Indian Territory and then was extended along the Mason-Dixon line in the East. Then, in 1906, the federal government launched a labor-intensive effort to eradicate the southern tick to prevent the transmission.[41] The US Department of Agriculture focused on killing off the carrier instead of on breeding animals with immunity or inoculating them. Tick eradication required the "dipping" of southern cattle in vats every two weeks from the spring to the fall, an effort that went on for nearly four decades. The animals ran through channels filled with an arsenic solution—the cattle would swim or flail to keep their heads above the chemicals. This was a dangerous process that could kill or injure animals due to the solution's toxicity.[42] Even after the threat of infection was reduced by the government's eradication efforts, a buffer zone along the Rio Grande in Texas was maintained to prevent reinfestation from Mexican livestock.[43] The methodological approach to the Texas tick amplified and exemplified the eugenic perspective on societal improvement and agricultural reform.

Nowhere in American society did this belief in genetic selection take root more firmly than in the agricultural bureaucratic networks of the federal and state governments and land-grant universities and agricultural colleges.[44] The American Breeders' Association provided the organizational support for these government and university officials, and through conferences and publications, it helped sustain the groupthink that undervalued animals possessing certain economic and ecological strengths, particularly in the South and West. Despite the goals of specialization and standardization that rationalized the primacy of British stock, eugenics manifested in an irrational, cultural reflex rooted in "master race" theories that called for the eradication and extermination of animals valuable to different topographies and climates, a reflex that also had the effect of redirecting American farmers' money away from Mexican breeders and exporters to British producers.

The Eugenics Movement and the Establishment of "Improved" Breeding Programs

Famous for his role in inventing the telephone, Alexander Graham Bell was also an ardent eugenicist and worked directly with Jordan in his capacity as chair of the American Breeders' Association's Committee on Eugenics to draft a document setting out the principles of eugenic ideology. The report presented to the association's 1908 conference was steeped in "master race" beliefs, revealing the racist hierarchies the members of the committee espoused. Jordan, Bell, and their colleagues were concerned about the disintegration of civilization due to the mixing of "superior" races with "inferior" races. But much of the report was devoted to breeding and race "purgation." They wondered aloud—using a racist pseudoscience as evidence—about the possibility of breeding a better people and, thus, creating a better social order.[45]

Following the submission of the report, Bell stood in front of the audience in Washington, DC, and commended the Committee on Eugenics' efforts. He declared to these like-minded eugenicists that the association needed to further consider the value of incestual breeding in humans through "consanguineous marriages," referencing the work that had been done on the so-called superior breeds of livestock. The animal example led him to wonder whether only inbreeding among "inferior" humans was a problem and to speculate perhaps that inbreeding among "superior" people would have a positive societal effect by, as in the case of superior breeds of animals, perpetuating the most desirable traits.[46]

Bell's oral contemplations highlighted the intersection between eugenicists' "master race" beliefs and the breeding techniques they used with nonhumans and wished to use with humans. Whether conducting research or making recommendations for plants and animals or humans, members of the association relied on the same fundamental principles of heredity.[47] They pushed a certain set of ideas about selection as a means of shaping the preferences and practices of practical breeders, researchers, and teachers, advocating for the elimination of unwanted traits through sterilization laws and negative eugenics and the transmission of desirable characteristics through positive eugenics or genetic selection, principles that in turn directly informed the creation of ideologies that were disseminated at the International.

Eugenic beliefs influenced many publications by professors and universities. The American Breeder's Association's journal articles frequently addressed issues dealing with racial purity, and Charles S. Plumb, professor of animal husbandry at Ohio State University, reworked Charles Davenport's eugenic ideas into a practical guide for livestock producers in *A Study of Farm Animals*.[48] Plumb's work and advocacy shaped the International and the college students who attended, and his adoption of Davenport's genetic understanding of improvement made clear the role of the association and eugenic thought in livestock improvement. Davenport ran the Eugenics Records Office in Cold Spring Harbor, and because of his relationships and sway with land-grant university agriculturalists, he directly influenced the national livestock improvement movement with his writings on the "complex nature of heredity." While Davenport noted genetic traits were passed to offspring from parents, which showed that genetic makeup was not random, he insisted that animal genotype did not amount to just a combination of the traits of both parents. Plumb quotes Davenport's assertion that an animal "for the most part . . . is not like either one of [the parents], nor is he like the combined." In fact, Davenport concludes, "the most that can be said is that the offspring *resembles* his parents, and that all his characters are to be found somewhere in his parentage," as livestock inherited characteristics from parents, grandparents, and even more distant generations. Improvement, Plumb insists, required farmers to manage the "persistence of heredity" in domesticated animals.[49]

Knowing the importance of heredity, however, did not oblige farmers to be prisoners of it. Instead, progressive breeders could shape the formation of family genetics to control the bodies of animals, navigating the ancestral tendencies of mating pairs to increase the transmission of positive traits and to eliminate negative qualities, including decreasing the likelihood of atavistic reversion or a predisposition for disease. Farmers adopted many practices to create food-producing animals that uniformly propagated certain body types.[50]

Plumb divided producers into two groups: "constructive" and "destructive." Plumb argued that destructive breeders held back agricultural improvement and thereby prevented the advancement of society. Nineteenth-century range producers, for example, fell into this category. They only intervened in genetic selection in their use of castration, a technique that entailed separating worthy males from inferior stock they would then cull.[51] On the other hand, the constructive breeder, according to Plumb, utilized the "right type" to uplift their herd.[52] The effects of both these types of breeding were not confined to an individual farm. For those in the livestock improvement movement, the advantages of "breeding up" or the disadvantages of scrub breeding improved or reduced, respectively, the productivity and well-being of society. Plumb argued that the use of either inferior or superior livestock "rippled" out into the broader agricultural community like a "stone thrown in the water, radiating out in still wider and wider circles." Consequently, producers, according to Plumb, had the responsibility to assist reform efforts by pairing animals with "harmonious rather than antagonistic qualities."[53]

Bakewell's ideas on breeding provided the foundation for Plumb and later agriculturalists. While many breeders in Bakewell's time bred animals for size alone, he rejected big animals as a meat type. He closely studied his animals' forms and proportions, which helped him maximize carcass parts associated with quality flesh, such as the loin and rump. These priorities led Bakewell to breed early maturing animals of uniform market design. His insistence that physiological and aesthetic traits, such as color pattern, were inheritable and could be controlled by the breeder also ran counter to the beliefs of many of his contemporaries. Bakewell maintained that these traits could be passed down to offspring by limiting the number of genetic possibilities through controlled breeding.[54]

Bakewell's ideas were persuasive to breeders, as evidenced in the familiar maxim "like begets like," and they became the foundation for pedigreed animals through which like animals were identified and closely bred.[55] Agriculturalists looked to Bakewell to better understand the development of purebred livestock and uniformity within breeds. Inbreeding and linebreeding became common tactics to increase genetic similarity in a herd. Through inbreeding, which limited potential outcomes in offspring, it became possible to standardize improved qualities, and that in turn contributed to establishing the breed as superior. Consistency resulted from eliminating the diversity of competing traits and thereby giving a statistical advantage to strong traits.

The packers' and professors' reverence for British producers was not limited to Bakewell. Shorthorn-breeder Thomas Bates's strict use of inbreeding also greatly influenced turn-of-the-century agriculturalists. Bates rejected old traditions of only "crossbreeding every now and then" to inject new life or traits into a herd, which helped avoid the weaknesses intensified by inbreeding.[56] In 1915, the *Journal of Heredity* featured Bates's careful evaluation and breeding of Shorthorns at the beginning of the nineteenth century. Like Bakewell, his successes in specialized breeding and genetic selection made him a transnational and transgenerational icon. In this reflection on the development of the Shorthorn breed, his livestock too garnered attention for being the foundational herd among Milking Shorthorns.[57] Bates focused on the production of heavy-milking cattle and developed a closely bred family of Shorthorns, shunning the introduction of "mongrel vigor" into his elite cattle, which would have compromised ancestral purity. Inferior animals introduced too many potential outcomes in their genetic pool. Consequently, purity in genetic composition amounted to status, to be sure, but also control.[58]

Famed breeders used different types of inbreeding, such as mating brother to sister, father to daughter, and mother to son. For F. R. Marshall, professor of animal husbandry at Ohio State University and official judge at the International, these matings were dangerous and undesirable.[59] He offered one of only a few dissenting opinions regarding these practices. Marshall's expertise on purebred animals and breeding—the US Department of Agriculture sent him to New Zealand and Australia on a mission to evaluate livestock types and import "promising breeds of sheep," and he facilitated the importation of the first Corriedale sheep to the United States in 1914—elevated his status at the International and among bureaucratic officials.[60] In light of his experiences with purebred animals and his examination of breeding outcomes, Marshall warned breeders of the defects and weaknesses created by limited genetic diversity. Interestingly, he drew on the human example in raising objections, citing the Bible and state laws that forbade cousins from mating. But he also referenced observable defects that arose from inbreeding livestock such as barrenness, predisposition to disease, and weak constitution or physical strength.[61]

For many purebred breeders, however, the risk of outcrossing animals outweighed the repercussions of inbreeding. They likened outcrossing to speculation because of the dilution or dissipation of genetics from strong heritage and believed that with each passing generation, as the offspring became more removed from famed ancestors, the animals' quality and purity declined. To minimize problems associated with inbreeding, they avoided the close matings of brother and sister or parent and offspring and opted for half brother and half sister. Typically, the half that the matings shared came from a famed stud sire or dam. Thus, farmers could still breed animals closely enough to improve uniformity. For these producers, Marshall noted, "the paramount question is not how much inbreeding is safe, but rather, how much outbreeding can be permitted?"[62]

Even with the use of close breeding, however, unwanted traits incidentally resurfaced in offspring; often atavistic traits emerged in descendants from foundational ancestors that had been deceased for decades. For example, the mating of two black Aberdeen-Angus cattle could result in a red calf. In this one hypothetical mating, a dormant trait emerged in the offspring, even though subsequent matings of the two parents might yield only black calves. Although this was an improbable occurrence, reformers urged breeders to eliminate such offspring because keeping and breeding animals with unwanted, atavistic traits increased the likelihood of them resurfacing in subsequent matings.[63]

Inbreeding, linebreeding, and close breeding helped farmers intensify traits through selection and mating decisions; however, to purge unwanted traits, agriculturalists also recommended negative selection practices. According to reformers, the progressive breeder bore the responsibility of eliminating animals, even purebred livestock, with unwanted features, and so they called for "severe culling" of undesirable livestock.[64] Improvement obliged breeders to sterilize and remove males and females from the farm that possessed inferior bodies to prevent the proliferation of those traits.

To manufacture better livestock, Marshall worked with his land-grant colleagues who created and ran the International to establish the basic principles of improved breeding. These reform-minded agriculturalists encouraged farmers to take into account the individual merit of the animal, the value of its offspring, and its ancestry. Marshall called these three components of improvement the "triple test." To help breeders, Marshall supplied his own scorecard for evaluating matings and the performance of genetic pairings.

By paying attention to the uniformity and productivity of offspring, Marshall suggested, farmers received valuable information about the ability of the sire and dam's genetic line to pass on traits. Marshall rated "uniformly siring good stock" high on the list of desirable qualities for parentage. An animal was "equally indebted" to each parent for its genetic makeup, and thus he advised farmers to give "greater value . . . to the pedigree of an animal whose sire and dam are both proved to have produced offspring of merit." Accordingly, Marshall advised producers to conservatively use sires or dams with few offspring to guard against overuse of unproven livestock.[65] Constructive animal husbandry required farmers to strike a balance between individual merit and the proof of quality offspring. At times, Marshall noted, the best-producing animals on a farm often lacked the aesthetic appeal of the show animal, but when farmers were assessing the parental productivity of a sire or dam, he argued they should give those characteristics a lot of weight. Marshall nevertheless advised breeders to find like parents to ensure advancement in physiological type and uniformity. And for the third test, improved breeders studied ancestors beyond the parents of any individual animal. When breeding, farmers not only needed to collect information but also had to understand the genetic impact of third-, fourth-, and fifth-generation ancestors. Distant ancestors, he cautioned, combined traits with other distant ancestors in the creation of offspring.[66]

Although these agriculturalists preferred British purebred livestock, they did not consider purebreds permanent or fixed but rather as both ancestrally pure and as under construction. This was the central paradox of the triple test: the irrational racialized bias was at odds with the rational economic imperative to continually work on and change the actual physical shapes of animal bodies. The uniformity and standardization that resulted from the interaction of these irrational and rational components of improvement rested on systematized incestual breeding. Constructive breeders utilized inbreeding and close breeding to shape livestock genotype and phenotype, narrowing the statistical variations of mating pairs. Purebreds, in other words, were the building blocks of the ideal animal, not established perfect animals ready for agricultural use.[67] Whether through mating choices or the elimination of unwanted stock, effective genetic selection amounted to control for farmers over the genetic and physiological makeup of livestock. By attempting to transform meat-producing animals, twentieth-century agriculturalists worked to fulfill Bakewell's goal of making machines of animals—machines that converted "vegetable products of the farm into animal products of greater value."[68]

The suitability of British purebreds to the industrial project reconfigured livestock ideologically as well; "improved" animals carried in their bodies iterations of biological technology. Pedigreed stock allowed breeders to manufacture their desired product by controlling potential undesirable outcomes—a process necessary for the creation of the ideal animal. Breeders treated their animals like engineered machines or duplications of technology. Rumor had it, for example, that Bakewell infected animals past their prime with parasites to prevent their genetics from being utilized by other producers—an infection that amounted to a grotesque form of intellectual property protection.

Better Sire Campaign

The erection of the Purebred Livestock Record Building in 1902 formalized the meatpackers' institutional and administrative support of the International's central goals. The breed associations that represented the different purebred animal varieties set the rules and regulations for breed shows; these associations governed their individual breeds and provided the requisite guarantees of authenticity and uniformity in their livestock.[69] By offering pedigreed proof and verifiable certification of farm animals, these societies were an essential cog in the reform movement. This certification process established the "purity" of the breeds—the central function of breed associations.

These "certificate[s] of purity" were fundamental to establishing, propagating, and expanding purebred animal husbandry.[70] The structure of the International itself normalized purebred thinking, making the guidelines of the breed societies necessary.

Within each breed, like Shorthorns in cattle, Cheviots in sheep, or Yorkshires in pigs, breed managers at the International divided classes based on animal sex and age.[71] The format for competition at the International relied on assurances from the associations that these animals authentically represented their breeds. The pedigree charts created by the registry allowed farmers to recall the sires, dams, grand sires, and grand dams of the animals they owned, which informed mating decisions. The identification—a unique number assigned to the animal—and name of the sire was placed on the top line of the chart and that of the dam on the bottom line. Many of the associations also included known data on the individual animals to give farmers a sense of potential productivity, which also helped farmers. By 1917, nearly all the major American purebred registries held their meetings at the International—thirty-one in total.[72]

Even though the group classes and the females at the International received acclaim and attention, the male animals—as sires and potential sires—received the pedigreed glory for generational improvement within breeds. For example, Ohio Chief, a pig bred by a well-known constructive porcine breeder in Ohio under the company name S. E. Morton and Company gained fame after winning at the first International in 1900.[73] Ohio Chief was a Duroc-Jersey boar, and he was featured by H. E. Browning of the *Swine World* in a retrospective in 1917 on this early period of Duroc-Jersey breeding that lauds the transgenerational impact this pig had on the breed. After winning the International, the owners returned to their farm to test his ability to pass on these elite genetics and physiological traits to his "get" (offspring). He quickly proved his "prepotency"; in 1904, the company fielded its show herd with "get" from Ohio Chief. Nearly all of the animals they showed at the St. Louis World's Fair in 1904 came from him, and Ohio Chief himself won again. Ohio Chief and his gets' success at the Louisiana Purchase Exposition resulted in the company being awarded the Premier Champion Breeder's Prize at the World's Fair, which the organizers of the hog show gave to the top breeder.[74]

Ohio Chief became a cornerstone in Duroc-Jersey pedigrees. A long line of highly sought-after and influential boars sired by Ohio Chief saturated the pedigrees of top Duroc-Jersey herds for the next two decades.[75] The emphasis on the male side of purebred pedigrees as well as the "get" classes that attributed animal quality and uniformity to the sire revealed a bias among breeders. The International established novel classes that emphasized the importance of sires in commercial farms too. For example, land-grant universities participated in the Demonstration in Mutton Improvement, the goal of which was to illustrate the generational value of purebred sires. In this class, schools bred purebred mutton rams with grade or crossbred ewes and compared results.[76] These ewes, the International specified, lacked the ideal form for meat production. But the ram was supposed to be the best example of a purebred meat sheep.[77] The offspring competed in different classes based on age and sex as in the other portions of the International, but unlike in the case of the purebred classes, the mutton competition normalized the principles of breeding up scrub livestock.

The prevalence of "inferior" animals on commercial farms in the United States posed a perennial problem for the livestock improvement movement. Plumb lamented that "a very large percentage of our breeders use only grade or scrub sires, which . . . explains why one sees so many inferior animals on American farms."[78] The International founders along with constructive breeders and the US Department of Agriculture developed a strategy that focused on the use of purebred males, not females, to address this limitation. The focus on males was a rational choice made by agriculturalists to improve the statistical and financial impact of improved animals. Agriculturalists would have preferred that purebred males and females filled the herds and flocks of each farm. But reality forced reformers to come up with strategies that maximized the benefits garnered from investments in improved livestock.

Each sire represented half of the genetic composition of a crop of calves, lambs, or piglets; a bull could breed nearly twenty cows, and that ratio increased for rams and boars. Spending additional capital on a single male rather than twenty or more females had the same statistical genetic outcome while allowing the farmer to reduce initial input costs. The use of these purebred sires reversed the degrading impact of scrub genetics by generationally "grading up" undesirable herds. The Department of Agriculture printed a diagram to assist in the breeding-up campaign that detailed the incremental, percentage-based benefit of reorienting a herd or flock's genetics with the use of a purebred sire (fig. 2.2). In the first mating on the left side, the fictional breeder mates a purebred sire with a grade female with half unknown traits and half known. After one generation of this mating, the grade animal improves to 75 percent known genetics, and by the fifth generation only a small fraction of the grade animal consists of scrub or inferior genetics.[79]

Figure 2.2. The influence of a purebred sire on improving grade stock over five generations. *Source:* Plumb, *A Study of Farm Animals.*

To increase the productivity and profitability of the farm, as De Loach noted, producers had two practical avenues: they could either produce more animals to increase the quantity of livestock in the United States or they could improve the level of productivity of each animal to ensure quality. Progressive agriculturalists encouraged the latter, arguing that greater economy was to be had in meat production if farmers utilized well-bred animals because of their ability to convert grain and roughages at high rates into quality meat instead of to bone, hide, and waste like the scrub.[80]

While the initial costs were higher for constructive breeders, these investments yielded increasing revenue. Conversely, the retrograde impact of inferior animals bred by destructive breeders undermined meat supply, and national food output and had a negative effect on their earnings.[81] De Loach argued that the income potential in marketing higher-grade animals with better yielding carcasses (fig. 2.3) outweighed increased input costs.[82] The market rewarded producers who sent purebred animals to the Union Stockyards. From 1895 to 1899, farmers made 20 percent more revenue on average because they sold improved animals. This price difference stemmed in part from increased consumer demands for better quality meat and in part from the greater value of the entire carcass due to a higher percentage of edible products per animal.[83]

When the US Department of Agriculture embarked on a "national crusade" to improve "Uncle Sam's livestock," the livestock improvement movement's goals became the official policy of the federal government. Concerned by the low number of prime and choice meat animals being sent to the Chicago market and the underperformance of dairy cattle, the department launched the "Better Sires—Better Stock" campaign in 1919. This campaign espoused the eradication of inferior sires and emphasized the ability of purebred sires to hasten improvement. The most notable participant in the program was President Woodrow Wilson. In 1920, Wilson registered his flock of sheep that grazed on the White House lawn. Countrywide, county agents enrolled producers who dedicated herds and flocks to the sole use of purebred sires. Ownership of those sires was not required; the farmer could rent, share, or borrow as well. In return for this commitment, producers received a certificate and a lithographed sign—a metal plate—advertising their farm's commitment to well-bred stock. Designed to be hung on a barn or at the farm's entrance, the sign read "Purebred sires exclusively used on this farm"; the sign was a source of pride and tool for advertising for the farmer and simultaneously promoted the Department of Agriculture's national crusade for better sires.[84]

Even though the campaign focused on males for strategic reasons, the livestock improvement movement did not discount the value of females. D. S. Burch of the Bureau of Animal Industry, a subdivision of the Department of Agriculture, unequivocally stated his desire for this effort to include all classes of animals, including females. He hoped to simply convince farmers to use a single purebred male and allow them to witness the results, which would instigate their participation in the livestock improvement

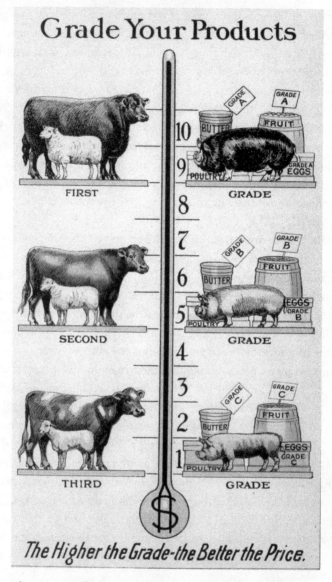

Figure 2.3. An illustration of the value of grading in livestock products. *Source:* De Loach, *Armour's Handbook of Agriculture.*

movement. Burch was convinced that this results-driven tactic—a prove-it-to-the-farmer approach—manifested in producer acceptance of well-bred animals. Constructive breeders witnessed a 50 percent increase in animal market value when they used a single purebred male, and they generated

40 percent more revenue, which Burch believed was all the evidence the farmer needed to continue herd uplift and become breeders of purebred females as well.[85]

Aware of farmer reluctance and price hurdles, the bureau instituted a series of county-level campaigns that allowed farmers to experiment with a purebred male. County agents hosted public pig-raising contests, dressed-carcass demonstrations, and auction exhibits. All of these events illustrated the gains made in feed efficiency, live and butchered weights, and market value. The most dramatic attempt to persuade breeders was the scrub-sire trials conducted in the 1920s, echoing the eugenic ideologies of animal criminality. These were an immediate hit—so much so that the bureau received over five hundred requests for the mock trial outline just shortly after it was first published. These trials mirrored the procedures of human trials and similarly cast the scrub as a societal problem. Thus, this court of animal justice included a judge, sheriff, prosecutor, defense attorney, jury, and the defendant—the scrub bull, boar, or ram. The case made by the prosecutor and the defense attorney highlighted the merit of purebred sires and the drawbacks of inferior animals.

Nearly all trials ended by declaring the scrub guilty of vagrancy and larceny, and the scrub's victims included his owner, his offspring, and the community. Following the conviction, the judge sentenced him to execution (fig. 2.4). Then, as outlined by the bureau, the scrub was removed, and the audience heard a gunshot. While the scrub was not necessarily killed on the spot, the theatrical use of gun left no observer wondering about the fate of the defendant. The bureau also encouraged ending these trials with a funeral procession and oration describing to the public the case against the scrub. The trials were followed by purebred sales, music, and a barbeque—in some cases the spectators ate the guilty party.

The scrub-sire trials were an immediate success, primarily in the southern states. However, simply convincing breeders of these advantages did not directly translate to purebred sire usage. The bureau and county agents also developed a series of purebred-sire distribution plans. To defray costs, cooperative units were organized to jointly purchase bulls, boars, and rams, and then the farmers who were members of the cooperative shared the sires. These cooperative bull associations and ram rings ameliorated cost, and the farmers retained breeding rights and thus the full benefits for their farms.[86]

Counties also established breeding associations and held county-endorsed and -managed sales to encourage the purchase and distribution of high-class

Figure 2.4. A scrub bull in court. *Source: Some Tested Methods for Livestock Improvement.*

sires. Perhaps one of the more creative strategies was developed with the tactical help and finances of railroad companies in several northern states. The Michigan Railroad Company and New York Central created scrub-sire exchanges. The companies sent trains filled with well-bred, pedigreed bulls throughout northern states. When they arrived at rural depots, the purebred bulls were exchanged for scrubs at sire swaps. Then the railroad companies loaded the so-called inferior males onto the train cars designated for scrubs, often referred to as the outlaw cars, and sent them to slaughter (fig. 2.5).[87] The International also joined in these efforts by creating new competitions that featured the value of breeding purebred males with inferior females.

Sire-centered breeding up programs were pragmatic approaches to improvement that contributed to achieving the federal government's and the International's broader goal of transforming the American meat supply. Not only was focusing on a single male more cost effective, but a single "top" fe-

Figure 2.5. A scrub bull entering the "outlaw car." *Source: Some Tested Methods for Livestock Improvement.*

male only birthed a single litter—cattle typically only having singles—whereas utilizing an "elite" male genetically impacted 50 percent of offspring on the entire farm in the first breeding year. Intact males' purpose on the farm was explicitly reproductive; they did not produce milk nor were reproductive males kept for slaughter. Because their value lay wholly in their reproductive merit, the farmer retained as few reproductive males as possible, using mating ratios to determine how many were needed to increase genotype and phenotype consistency and to enlarge the genetic footprint of well-bred males while minimizing input costs.

CHAPTER 3

Recreating the Animal Body

Whatever the animal kingdom can afford for our food or clothing, for our tools, weapons, or ornaments—whatever the lower creation can contribute to our wants, our comforts, our passions, or our pride, that we sternly exact and take at all cost to the producers. No creature is too bulky or formidable for man's destructive energies—none too minute and insignificant for his keen detection and skill of capture. It was ordained from the beginning that we should be the masters and subduers of all inferior animals.

Richard Owen, *Lectures on the Results of the Exhibition Delivered before the Society of Arts, Manufactures, and Commerce,* 1851

Only twenty years old at the first International, Fred Hartman of Fincastle, Indiana, carried the energy of progressive agriculture within him. He traveled the United States showing his well-bred sheep, ending his season at the International.[1] By that time, Hartman's animals knew the show ring and anticipated the movements and rigors of the showman and the competition and no longer behaved like the flighty, scared sheep in the slaughter pens on the same grounds. At the 1900 International, as Hartman held his ewe, Beatrice, under the chin with his left hand, the judge approached her to feel her consumable parts.[2] Beatrice, accustomed to the close inspection, walked around the ring and stood for evaluation. Many farmers and producers saw her pictures and accomplishments in national journals; judges praised her, and Hartman's contemporaries recognized Beatrice for her long record of winning top prizes.

Hartman avidly promoted purebred livestock, especially the Cheviot breed, which originated in Great Britain. He argued that Cheviots possessed qualities necessary to the general improvement of a producer's farm value. His flock, Hartman bragged, contained "the best that money could buy or science could produce." For the modern farmer, science was critical, as it correlated to improved genetic selection and the physical formation of the ideal animal. Under the farm name of Maple Grove Cheviots, Hartman's sheep won 378 premiums and 212 first place honors between 1900 and 1903. His sheep excelled in "uniformity of type . . . shortness of leg and quality of highest type."[3]

Turn-of-the-century judges linked uniformity and shortness of legs, or compactness, to productivity in meat-producing livestock. By focusing on distinct physiological designs, not just aesthetic traits like color, agriculturalists connected animal specialization with farm specialization. This version of efficiency espoused at the International revolved around the animal's place in and contribution to the industrial sequence. This sequence started with the specialized corn farmers who harvested their crop and sold it off the farm to livestock producers, sometimes in different regions of the country. The animals then converted those grains combined with foliage into meat, milk, bone, and fiber. In the modern animal, the proportion of these, meat to milk, for example, dictated its function. The deliberate reorientation of animal function manifested in the reconstruction of the animal body. Once the animals had effectively converted grain to meat, buyers transported animals on the vast network of emerging rail lines to major disassembly factories. The omnipresent Chicago meatpackers dominated this part of the sequence. After harvesting, meatpackers processed the animals' commodities, including by-products, for delivery and consumption.

An enduring paradox in twentieth-century agriculture was that as food output and agricultural yields grew and gross revenues increased, farmer profitability remained unstable. Efficiency in corn-to-animal food production required farmers to operate on tight margins, pushing them to increase farm size in order to remain economically viable. All the while, animal work and the animal body created profit for agricultural stakeholders at every point of this sequence, providing opportunities for those who were neither a corn producer nor livestock feeder. As the center of a vast network of agricultural improvement projects, the International incentivized and normalized this concept of efficiency by reshaping animal bodies. Although seemingly benign, agricultural efficiency was part of an explicit consolidation of American agriculture into a singular system that positioned the animal as an industrial being, showing little concern for the ecological and regional demands or for the role of animals in the places they were raised.

Reformers used the goal of industrial efficiency to justify the central thrust of livestock improvement: biotic control. By altering the meaning of animality, these reformers changed what cows, sheep, and pigs were. These changes to their bodies, genetics, and routines also changed how animals procreated, their relationships and experiences with their offspring, and their interactions with the farmers and fields around them. For example, farmers who prioritized and selected primarily for muscle consequently created animals that often had difficulty breeding, birthing, and raising their offspring, causing serious health problems for mothers and a more coercive relationship between farmer and animal.

Specialization was the key to efficiency according to both the professors and the packers. Medical doctor and animal husbandry expert Manly Miles, who worked for Massachusetts Agricultural College and Michigan State University, explained, "a high degree of excellence in two or more [qualities] cannot be obtained in the same animal" but "an extraordinary development of a single character" can be.[4] To encourage the creation of single-purpose animals, university and experiment station researchers set out to define ideal forms. But while the packers and professors agreed on the types and purposes of animals, they linked livestock specialization to animal efficiency for different reasons. The public-funded researchers were concerned with farm yield, farmer and rancher revenue, and gross food output. By increasing their carrying capacity, farmers theoretically would glean more revenue per acre farmed and consumers would have more food available to them per acre farmed. Packers, on the other hand, pushed for this type of specialization to encourage product standardization and increase the animals' slaughter value and thereby squeeze more profit from each animal body and further penetrate urban markets: livestock needed to be "subjected to careful standardization from beginning to end, in order that the best product and highest prices may be obtained," as R. J. H. De Loach put it.[5] Packers and professors thus collapsed the potential differences between specialization and standardization in their account of what efficiency meant for American agriculture by prioritizing genetic and physiological similarity over animal adaptability and biological diversity.

Altering the bodies of animals, literally reworking biological and so-called natural beings into an industrial mold, was driven by a public-private partnership that focused on redefining animality itself as a sort of public works project. This effort to rebuild animals through coercion and violence using techniques such as forced breeding, sterilization, culling, and extermination amounted to the construction of micromachines. By aggregating these animal machines from each farm, packers and professors created a national public infrastructure in the bodies of animals that directly influenced the trajectory of the twentieth-century industrial farm. As the hub for the national livestock improvement movement, the International incentivized the altering of physical forms to meet these production goals. Exposition judges prioritized muscularity, which corresponded to valuable carcasses, to encourage the reconfiguration of animals based on the goals of specialization and standardization. At the International, the potential outcome of the carcass dictated a farmer's breed choice and modern animal form.

The Intellectual Foundations of Animal Specialization

By way of his research, teaching, and advocacy, Charles S. Plumb served as the central figure connecting university campuses and experiment-station research to the International. His animal husbandry articles and books were widely distributed and used as essential texts in secondary and postsecondary schools as well as by practical farmers. Many of his books were translated for use at foreign universities, and *Types and Breeds of Farm Animals* (1906) became one of the two most influential husbandry texts in the United States over the first decades of the twentieth century. His commitment to establishing the physiological standards of modern livestock led him to take on the task of coaching Ohio State University's collegiate livestock judging team at the International. His status as a preeminent historian, prolific author, and improved livestock reformer earned him many awards, including induction into the Saddle and Sirloin Club Hall of Fame, a citation for distinguished service from the French government, and honorary doctorates from Massachusetts Agricultural College, Purdue University, and Ohio State University.[6]

Plumb anticipated the packers' and professors' work at the International. Selecting animals based on body type, for Plumb, required manufacturing animals' bodies in the same way an architect designed the ideal home. To erect a great structure, he argued, the architect first needed to have a plan, resulting from studying form and function, that would provide both a foundational understanding of proper construction and inspire a desire to do better.[7] Not all breeds of cattle produced milk and beef equally, and thus the first step in building the ideal breeding program was to identify the breed of cattle, sheep, or hog that produced a single commodity efficiently. The farmer selecting the proper breed differed little from the architect using the correct materials to build a sturdy home.[8]

Plumb's work detailing livestock type and evaluation was part of a broader academic and professional movement. John A. Craig, professor of animal husbandry at the University of Wisconsin, developed the first classroom adaptation of livestock judging instruction in 1892, which he turned into a book titled *Judging Live Stock* that was published in 1901. American and Canadian colleges treated both Craig's and Plumb's works as authoritative texts until Carl Warren Gay, professor of animal industry at the University of Pennsylvania, released *The Principles and Practice of Judging Live-Stock* in 1914. This book details the process of livestock evaluation and the importance of animals having specific purposes and deploys the phrase "the animal machine." For Gay, standardizing physiological forms in livestock was a requisite step to achieving the broader goal of mechanizing biological beings: the best type of domesticated animal, he argues, is "the most efficient machine for making the greatest return, in its specific product, on the raw material consumed." By transforming raw materials such as corn and rough forages "not available to man in their present form into animal food products," Gay explains, "the animal machine serves a most important economic purpose."[9] Gay cites British zoologist Richard Owen to justify not only the extraction of animal products and by-products after slaughter but also the subjugation and redirection of animal forms. In a lecture Owen delivered before the Society of Arts in 1851, he declared to the crowd that humans had been ordained to conquer all animals, whether big or small. The "slay[ing], subjugat[ing], and modify[ing]" of animals was within the domain of national interest.[10]

Following in Gay's footsteps, Robert S. Curtis, associate chief in the Animal Industry Division of the North Carolina Agricultural Experiment Station and Extension Service in Raleigh, published *The Fundamentals of Live Stock Judging and Selection* in 1920, which also went through several editions. Curtis reaffirms purpose-oriented breeding and clarifies the two central objectives of modern livestock production: raising animals geared toward a single purpose, which allowed farmers to produce the food necessary to feed a growing class of urban Americans and improve farmer income, and raising animals whose body type complemented function. Cattle, for example, bred for dairy needed distinctly different bodies from those bred for beef, and farmers had to reconfigure the actual shape of their livestock to better fit their purpose. Curtis liberally uses words that highlighted the key aims of improvement, referring to how the "modern judge," in contrast to his nineteenth-century counterpart, held the "ideal" animal in mind when surveying classes—a type of livestock consistent in "form" or "type" that correlated to "purpose."[11] This litany of constructive-breeding catchwords linked this network of agricultural reformers to a singular goal—directing the market performance of animals by establishing types.

In his audacious recommendations regarding animal purpose and type, Curtis undermines the alleged value and utility of dual-purpose animals by pointing to their yield drag, an injury to output that resulted from the splitting of consumable calories and the breeding of animals with body types intended to serve multiple purposes.[12] Echoing Plumb's metaphors of architectural design, Curtis also insists that British purebreds served as the building blocks of the single-purpose regime: establishing production goals and learning about the different types of animals available would, he maintains, improve "conformity of purpose."[13] Breeds, Curtis writes, "have been developed along specialized lines for performing definite kinds of work" and "practically all of them may be grouped into certain standard types."[14] Purebred animals and their breed proclivities provided farmers choices by establishing a transgenerational pattern for certain feed, soil, climatic, and work needs and possibilities.

These canonical texts in livestock evaluation directly influenced the preferences of International judges and constructive breeders and the education of agricultural students. Paired with the proceedings of the International, they defined the procedures for the selection of modern livestock. To be sure, the International offered incentives to encourage farmers to change their practices and provided practical displays of improved livestock, but academics did the professional and grassroots work of labeling, describing, and disseminating information about form and function.

In the beef industry, Shorthorn, Hereford, and Aberdeen-Angus cattle became the most prominent breeds in the early twentieth century. All three breeds originated in Britain, and breeders and meatpackers alike fancied them because they produced the choicest meat. Shorthorns technically were dual-purpose and had a strong genetic line inclined toward milk production. But although both lines came from the breed, class setup and judge preferences required breeders to clearly define and separate meat-producing livestock, including Shorthorns, from animals geared toward dairy production. Judge John Lewis, for example, commented on the Beef Shorthorn's lack of prominent hooks. The hooks or the hipbones observable on the topline became less visually noticeable when meat livestock possessed a carcass with the desired condition or fat. Dairy cattle in particular had prominent hooks (fig. 3.1), but so too did beef cattle with limited muscle shape or fat cover, as seen in range cattle.[15] Judges, therefore, relied on hooks to distinguish the purebred Beef Shorthorns from milking cattle and range animals. Judges utilized these visual cues to evaluate the probable productivity of carcasses and economic competitiveness of meat-producing animals.[16]

Shorthorns at the time were the most common breed in English-speaking countries around the world and, reflecting the intersection of international politics and animals, went by the empire-laden nickname "the universal

Figure 3.1. Mamie's Minnie, a milking Shorthorn. *Source:* Lloyd-Jones, "What Is a Breed?"

intruder." The name of the breed came from the shape and size of the horns. In contrast to Longhorns, Shorthorns had a medium-sized horn with a curved shape positioned forward and slightly downward. The smaller horns helped with transportation and handling, and their physical build made them useful for meat production. The meat-type Shorthorns possessed wide, strong backs and large bodies, and they matured quickly.[17]

Sheep produced wool in addition to meat, and certain breeds garnered recognition solely for their high-quality fleeces, including Merinos, Lincolns, Cotswolds, and Leicesters. The International, however, did not feature the wool breeds; instead, the show encouraged the development of sheep specialized in the production of mutton. Medium-wool sheep, including Shropshires, Southdowns, Hampshires, and Cheviots, were classified as mutton types. Thus, the class setup at the International, which included only breeds tailored toward meat production, forced breeders to buy and raise sheep from this category.[18]

American hogs primarily came from southern Europe, Great Britain, and China, and purebred pigs shown at the International were more genetically diverse than its cattle or sheep. Two types of pigs were prevalent in the United States: lard hogs and bacon hogs.[19] In the early twentieth century, lard hogs were more popular than bacon hogs among American breeders. Compact in form with bigger tops, more depth of body, and larger hams, lard hogs produced superior cuts of meat and more fat across their backs, while the bacon type possessed a narrower and longer body that was accompanied by a deep side, making it higher producing for the specific cut of bacon.

Breeders in Ohio developed the Poland China, for example, from crossing Russian, Byfield, Big China, Irish Grazier, and Berkshire hogs. This mixture in breeding yielded diversity in form in the late nineteenth century, but as the breed developed more toward a uniform type, agriculturalists came to categorize them as lard hogs. They accumulated fat with ease and possessed large hams and tops. In the corn-producing states, Poland China hogs had a reputation for being "pork-packing machines."[20]

Husbandry texts cast modern agriculturalists as animal "subduers" and created a taxonomy of priorities for farm specialization. The "animal machine" was central to these reform efforts; modern livestock was vital to the broader industrial sequence in food production. Selecting a breed that correlated to the farmers' commercial purpose was only the first step in this taxonomy; breed simply served as a foundational tool in building the modern animal.

Livestock Shows and Animal Form

At the St. Louis World's Fair in 1904, the agricultural exhibits still included range cattle of the West that little resembled the "improved" animals at the International. Kansas State Agricultural College's display—the "red bovine mastodon"—offered observers the most visceral juxtaposition between the goals of the International and the older-style market animal. The older-style steer named Sampson was massive and drew crowds of spectators and both positive and negative attention. He was four years old, weighed thirty-five hundred pounds, and was over six feet tall and nine feet long (fig. 3.2). His height and weight were remarkable compared to nearby purebred cattle and the human spectators.[21]

Figure 3.2. Sampson. *Source:* Irwin, "Agricultural Events at the 1904 St. Louis World's Fair."

The awe Sampson's sheer size inspired was a product of nineteenth-century notions of excellence. Agricultural journals and fairs heralded giant animals and plants as sources of amusement, and improvement societies latched on to these monstrous beings as examples of advancement for culti-vated biological products, linking superlative agricultural products—those that bent or even broke the limits of nature such as freakishly large strawber-ries and tomatoes and astonishingly big animal bodies and carcasses—to human achievement. This assumed correlation between the spectacle of gross size and the spread of agronomic improvement traversed the Atlantic Ocean, connecting British husbandry and its replication in other places via colonial domination and extraction to American territorial expansion and agricultural development. In fact, in the view of many American improvement societies, large fruits, vegetables, and animals validated notions of white superiority and ingenuity that accompanied the physical expansion of the British Empire and American settler colonialism.

To further this cause in the United States, state fairs and agricultural ex-hibitions featured monstrous plants and animals as standards of agricultural advancement; these show-ring spectacles tied the developing norms at Amer-ican fairs to traditions established at British livestock expositions.[22] The Fat

Stock Show in the Inner-State Industrial Exposition Building organized by the Illinois State Board of Agriculture in 1878, for example, was guided by the "bigger is better" dogma characteristic of nineteenth-century exhibitions in the United States and Great Britain. These show animals exhibited impractical bodies, which reflected a broader trend among producers who preferred old, extremely heavy livestock.

Exhibitors commonly showed four- or five-year-old cattle that weighed as much as twenty-five hundred pounds.[23] In fact, the animals' physical features sometimes included large, awkward chunks of fat on foreheads and legs; their limited ambulatory abilities owing to this fat even necessitated special wagons at British shows to transport them to the show ring.[24] The biological limitations of these animals gave critics of nineteenth-century livestock ample ammunition; the absurd size and fat cover of show stock, they noted, reduced health and productivity, which had ruinous effects on agriculture and by extension the food source of the nation.

In Britain, the stated goal for showing sheep, cattle, and swine was to improve meat production. However, the reality of purebred livestock production, especially as the ego-laden show ring pushed producers to excess, conflicted with this rhetorical aim. Breeders, for example, instructed painters producing portraits to embellish the height of the animals and to shrink the size of the showmen. British producers and showmen sought to create an animal that dwarfed the livestock of their chief competitors; aristocrats touted these oversized and "noble" animals as a sort of biological representation of their own elite status.[25] Buying or selling these high-priced animals amounted to conspicuous consumption that utterly discounted the agricultural value of animals and instead signaled the aristocratic practice of collecting precious items inaccessible to the average farmer.

Before the Fat Stock Show ended its run in 1893, it inaugurated a competition that reflected the production concerns of an emerging class of land-grant university researchers who urged farmers to calculate cost and yield to define what efficiency meant. In this cost of production competition, the farmer whose animals gained the most weight per day at the lowest relative cost won the prize. Farmers documented feed input and weight data for their animals at different ages, and this information allowed agriculturalists to calculate the amount of weight gained per pound of feed given to steers. Younger steers, it turned out, converted feed more efficiently. For the average steer at the show, gaining one pound cost $3.21 in his first twelve months, when he gained the most weight. The cost rose to $4.56 in his second year, and in the third year the cost to produce a pound was $7.60, when the steer gained the least weight. Steers that were provided good quality feed typically made remarkable gains through twenty-four months; as the animal aged, the daily gains decreased, and the cost soared.

This cost-of-production analysis defined efficiency for reformers. Elliot W. Stewart, editor at the *National Live-Stock Journal*, analyzed this data and used the information to advocate for encouraging farmers to produce early-maturing livestock. In his popular text *Feeding Animals: A Practical Work upon the Laws of Animal Growth*, he demonstrates that steers surpassed maximum productive efficiency before twenty-four months and that older steers' value diminished relative to cost. This model for the efficient production of livestock drove reformers a decade later at the International to prioritize early maturity. But in the 1880s, Stewart admitted that this model was only in its "infancy"; judges and showmen disregarded "the most economical beef animal" for "the heaviest beef animal."[26]

The ideal steer at the International little resembled these nineteenth-century livestock. Packers and professors initiated a shift to make animals more economically efficient by encouraging farmers to achieve a finished or butcher weight for their livestock in a shorter period of time, compelling them to take not just animal ancestry but also function into account when it came to mating.[27] Decreased finish weight had a compounding impact on quality and condition; thus, there was a strong correlation between weight and value. Top-quality steers were rated as prime, and they weighed between twelve hundred and sixteen hundred pounds. Lower-quality classes, including both choice and good steers, had a similar weight range. Common rough steers weighed between nine hundred and twelve hundred pounds. The lower end weights fluctuated, but the ceiling on top weights on all grades were dramatically less than Sampson at thirty-five hundred pounds.

A shift occurred in favor of early-maturing animals with a smaller skeletal frame. Ultimately, older, bigger animals resulted in high production costs and, pound-for-pound, were worth less on the market.[28] Plumb contended that the new type of meat animal was a steer between eighteen and twenty-four months old, and James Poole, livestock market expert who wrote for the *Breeder's Gazette* and articulated the packer perspective, lauded the International for eliminating aged steers. Judges' preferences, in addition, discouraged showmen from exhibiting two-year-old steers, even though they were not considered aged, in effect creating a de facto ban on cattle two years old or older. They selected grand champions almost exclusively from the yearling

Figure 3.3. Erwin C, grand champion Aberdeen-Angus bull in 1913, W. A. McHenry, Denison, Iowa. *Source: A Review of the International Live Stock Exposition*, 1913.

classes. By 1919, these tastes had become normalized as a result of the standards set by the International judges, and Poole declared that "in recent years by common consent the yearling has held undisputed sway."[29]

In the cattle industry, dairy cattle had very different bodies from beef cattle. Dairy cows had long, wedge-shaped bodies and often had thin necks, prominent hip bones, and full, square udders. Conversely, beef cattle were short and blocky. The beef animal had "fill" (muscle and fat) in areas the dairy one did not. Beef cattle, ideally, had broad backs, prominent loins, and full rumps. The 1913 Grand Champion Aberdeen-Angus Bull, for example, featured many of the ideal traits of the modern animal (see fig. 3.3). He was deep in his rump and rib—an indication that he carried flesh in great amounts where the highest and most desirable cuts of meat were located on the animal's body—and possessed condition (fat) evenly across his body, giving it a smooth appearance.

Even more, reformers believed this bull demonstrated masculine sex character, which provided visual cues that he would be an effective breeder. The 1918 Shorthorn bull champion exemplified these features.[30] Lord Rhybon had "depth of body," which he carried down his broad back and deep into his lower quarter.[31]

To standardize these characteristics, the organizers of the International depended on the credibility, knowledge, and reputation of the judges who placed the cattle, sheep, and swine.[32] The International needed the participants, urban spectators, and foreign observers to recognize that even though the judges used their own discretion in prioritizing observable traits, they all had the same core ideals. Otherwise, the show ring would seem irrational or unscientific. To further improve the reputation of the show, the International featured judges from countries with similar tastes for purebred animals, such as Britain, Canada, and Argentina, and the organizers parlayed these judges' compliments and positive observations into advertisements for the show. At the first International, for example, J. B. Ellis of Walsingham, England, judge of the fat cattle classes, argued that it would be difficult to compare English shows with the International, but "Shorthorns, Polled Angus, Galloway, Red Polled, and . . . the Polled Durham . . . made me loath to confess that the best exhibits in these breeds could not be excelled anywhere."[33]

In 1916, the International expanded its foreign delegation to include Argentine cattlemen. The Honorable Carlos M. Duggan of Buenos Aires, Argentina, judged the grade and crossbred class and champion steer competition. The International founders wined and dined the Argentine delegation, which included Ambassador Rómulo Sebastián Naón, at the famous Saddle and Sirloin Club. The Argentines repaid their gracious hosts with rousing appreciation of American livestock. Judge Duggan exclaimed, "I consider the grand champion steer, California Favorite, the best I have ever seen and honestly think he would be a winner at any show in the world." The steer's "great evenness, quality and wealth of flesh could not be beaten and to sum him up I would say that I think the most critical judge would find it a tough job to pick a fault in him." Duggan went on to rank the International first in the world among all shows.[34]

Each year, the judges sorted through the cattle, sheep, and swine classes to identify desirable traits among the animals that they then used to place the classes.[35] To be sure, no animal possessed the perfect form. Judges often had the responsibility of evaluating livestock with several undesirable or less desirable traits. Good judges nevertheless kept their preferences in mind as they examined each animal. They identified the most desirable parts and, comparing those parts to the animals in the ring, determined which animal possessed the most complete set of these traits. This "balance of points" process required deeper thought and consideration than any other aspect of livestock evaluation. As Curtis complained, many animals in the show ring "differ greatly in merits and faults" both as a whole and "in the correlation of parts." Hypothetically, he argued, if all animals differed in their same component parts and if those parts possessed a fixed value related to function, then judging would be simple. But that was never the case. The variation among animals in merits and faults and the location of those differing characteristics on the animals required judges to keep in mind a set of priorities. In the end, balancing points, for Curtis, obliged the judge to find the "relative value" of the animals based on "principles fixed entirely on utility requirements and the comparative value of correlated parts or units." Thus, for example, when a judge evaluated a class with two steers with several faults, the judge needed to know whether "a low back, scantily covered with flesh" in a steer represented a larger demerit than "one with a drooping rump, thin thighs, and high or open twist" based on the probable performance of the carcass.[36]

To help with this process, Gay provided an "index" that gave breeders, buyers, and judges a sense of market performance for each animal. Ideal meat animals possessed "broad, flat backs . . . low set, broad, deep, and . . . thick-fleshed," while inferior ones had long, narrow bodies and frames. Gay used these relationships of parts, whether narrow and shallow or deep and broad, to construct a useful, albeit clunky, dictum, what he called "the law of correlation," for judges: "As a rule, longitudinal dimensions of all parts are alike long or short and are inversely related to transverse and perpendicular dimensions." Gay's law directly linked long vertical and horizontal features, like leg and body length when viewed from the side, to narrow perpendicular traits, like shallowness of rib or thinness of rump when viewed from behind. Conversely, shortness of leg and body corresponded to a wider and thicker animal. Gay argued that this law applied to livestock in general: "Milk and beef . . . are opposed to each other by this same law."[37]

To assist in correlating "form to function" in making selection decisions, judges and breeders used two broad approaches. The first was the analytical approach, called "scorecard judging," in which judges rated each animal individually by prioritizing certain parts. After scoring livestock by their constituent parts, they added all the points from each section on the scorecard to find a total. The highest point total won the class. Scorecard judging was a

SCORE CARD FOR BEEF CATTLE.

GENERAL APPEARANCE—40 Points.   Perfect score,

Weight: score according to age   6
Form: straight topline and underline; deep, broad, low set,
    stylish   10
Quality: firm handling, hair fine; pliable skin; dense bone;
    evenly fleshed   10
Condition: deep, even covering of firm flesh, especially in
    regions of valuable cuts   10
Temperament: lymphatic, inclined to fatten   4

HEAD AND NECK—7 Points.

Muzzle: broad; mouth large; jaw wide; nostrils large . .   1
Eyes: large, clear, placid   1
Face: short, quiet expression   1
Forehead: broad, full   1
Ears: medium size, fine texture   1
Horns: fine texture, oval, medium size   1
Neck: thick, short; throat clean   1

FOREQUARTERS—8 Points.

Shoulder vein: full   2
Shoulder: covered with flesh, compact on top, smooth . .   2
Brisket: advanced, breast wide   1
Dewlap: skin not too loose and drooping   1
Legs: straight, short; arm full; shank fine, smooth . .   2

BODY—32 Points.

Chest: full, deep, wide; girth large; crops full   4
Ribs: long, arched, thickly fleshed   8
Back: broad, straight, smooth, even   10
Loin: thick, broad   8
Flank: full, even with underline   2

HINDQUARTERS—13 Points.

Hips: smoothly covered; distance apart in proportion with
    other parts   2
Rump: long, wide, even, tail head smooth, not patchy . .   2
Pin-bones: not prominent, far apart   1
Thighs: full, deep, wide   2
Twist: deep, plump   2
Purse: full, indicating fleshiness   2
Legs: straight, short, shank fine, smooth   2

        Total   100

Figure 3.4. Beef cattle scorecard. *Source:* Robert S. Curtis, *The Fundamentals of Live Stock Judging and Selection.*

means to an end; this approach provided the best educational outcome for students by requiring them to learn the various parts of the animals and the primary values of those parts (fig. 3.4).[38] Scorecard evaluation left "a mental impression of the ideal."[39] The second method was comparative one; practicality often forced judges to use this approach because of the imperfections in each animal. Thus, judges assessed characteristics in relation to other animals in the same class.[40] Even judges or educators who preferred the first approach often resorted to the second to settle disputes over their own preferences as the classes proceeded. To prevent conflict and enable judges to communicate their preferences and priorities, the International required them to provide reasons for their placings. At smaller shows, judges would sometimes simply talk to the audience from the ring, but at the International many judges published explanations or general observations regarding the classes they evaluated.[41]

In the first two decades of the International, the grand champion animals underwent remarkable physiological changes, which effectively demonstrated the influence of both the judges and the show ring through awards and penalties on livestock producers' priorities in animal breeding and culling choices. The physiological differences between Ruberta (fig. 3.5), the junior champion Shorthorn female in 1900, and Goldie's Ruby (fig. 3.6), the winner in 1918, clearly illustrate these changes. Ruberta had a smooth-made body with even flesh and a nicely laid shoulder; she did not possess the prominent hooks of the dairy cow.[42] She had a deep but not noticeably wide body from

Figure 3.5. Ruberta, Robbins and Sons, Horace, Indiana. *Source: Review of the First International Live Stock Exposition*, 1900.

Figure 3.6. Goldie's Ruby, Reynolds Bros., Lodi, Wisconsin. *Source: A Review of the International Live Stock Exposition*, 1918.

front to rear; she was a type of cow that had great balance and evenness. Goldie's Ruby, by contrast, demonstrated many qualities of the improved animal. Before she was selected as grand champion Shorthorn cow in 1918, she had placed first in a category for cows or heifers under two years old, having been deemed to embody the qualities that had launched the baby beef fad, meaning her body type was not as tall, big, or upstanding as the thinner-made range cow or even the typical Beef Shorthorn of previous decades.[43] She was much shorter and thicker than Ruberta, and carried an extreme amount of muscle and condition. She had a thick, broad back and round, deep rib shape. In fact, in the picture advertising her success, her chest, fore flank, rib, and rear flank barely cleared the straw bedding in which she stood—a clear illustration of what distinguished the beef-producing cow from both range and dairy cattle.[44]

By 1918, Alvin H. Sanders, chief International organizer and promotor, saw this shift as a linear progression and believed the Shorthorn breed, along with the International, had finally reached a place "free from fads." He had complained in the past about fads, follies, and fancies limiting or even prejudicing qualities in animals that prevented the march toward progress in animal husbandry. Sanders credited the "sanity of procedure" and abundance of "valuations" for the arrival of this "golden age." The International helped standardize traits and create a "rational," "scientific approach" to the reproduction and the manufacturing of the country's food source.[45]

But judges often found themselves faced with the problem of having to decide which physical traits to rank above others, the aesthetic ones that purebred associations touted or the pragmatic ones associated with muscling and carcass yields that the packers pushed. No judge or commentator ever resolved this conflict. Sanders likewise oscillated on the importance of breed standards and the visual appearance of the animal's head, but he concluded that the most important evaluative measure should be what he called "probable efficiency," but to embrace efficiency often meant having to overlook the aesthetic requirements of purebred animals.[46] For progressive agriculturalists, rib and rump shape and width of top on the animal correlated to higher yielding carcasses and, at times, that meant that judges had to ignore aesthetic qualities. Sanders warned that putting too much stock in aesthetic qualities prevented the transformation of cattle, sheep, and swine.

Despite this conflict between aesthetic and market-oriented physical features, the livestock improvement movement became a "crusade" to reshape animals' bodies, and the enthusiasm, beliefs, and missionary zeal of agriculturalists assumed religious overtones. R. R. Benson of Arizona, a contributor to the *Shorthorn in America*, argued that good judges devoted to modern agriculture and science were "prophets" of a new age who were often not recognized as such by average farmers because their standards for animal form seemed unusual. These "prophets" could see in the bodies and blood (genetics) of animals the short-term value for the farmer but also the idealized animal needed to improve agriculture.[47]

Standardization and Carcass Performance

To analyze the exact relationship between animal form and market performance, university researchers developed land-grant curricula and International competitions centered on animal carcass and meat cuts. In January 1907, Plumb addressed this functional need when speaking to a large gathering of agriculturalists that included Charles Davenport, Charles F. Curtiss, and F. R. Marshall at the American Breeder's Association's annual meeting at Ohio State University. In addition to presiding over the annual meeting, Davenport served as the secretary for the animal section of the conference, and he invited Plumb to lecture on the increased importance of prioritizing commercial value in animal selection.[48]

Plumb began his lecture by discussing the essential role of land-grant universities in the educative process, emphasizing faculty training and up-to-date facilities, and described Craig as the foremost leader in the academic community in the development of curricula on livestock judging standards for land-grant institutions. Then he shifted to the larger issue of animal selection that the International was concerned with. For Plumb, the foundational knowledge required for good animal judgement was the understanding of types.[49] Once students understood the broader function, ideal, or goal related to the general purpose and performance of the animal at the macro level, then the next step was for them to learn the constituent parts of each species and the specific utility of those parts in constructing the general physiological makeup of livestock. For example, students needed to learn how to analyze the most important cuts of meat on a beef cow.[50] In a textbook intended for students, Plumb illustrates beef animal parts, including the loin, rump, and ribs (fig. 3.7), and their retail value. Not all steers weighing the same amount were worth the same. Plumb wanted farmers to see that the distribution of that weight mattered and that the location of muscle and fat impacted the market value of the animal. The back contained the most valuable cuts of meat; the sirloin and porterhouse topped the chart in retail value.[51]

Curtis added to Plumb's work by providing more thorough diagrams and illustrations. He believed that "capacity" offered husbandmen a useful conceptual tool for modern selection. The core of the animal body needed to be like that of a barrel, ship, or wagon, rotund and full of space for blood flow to aid health, constitution, reproductive vigor, and, even more, the production of good-quality meat. Using capacity, shape, and volume as a gauge for evaluating the chest, guts, rump, or back of a steer meant looking for "a beef animal . . . [with] power to consume feed and convert it into proper material for body maintenance and development." This beef animal lent itself, according to Curtis, to a higher percentage of dressed carcass. He argued that if every animal just gained an additional one pound of quality meat, the American consumer could see as much as 173 million pounds more in edible product. These "square, low set, deep, broad in the body, compact and smooth" steers possessed a "large amount of weight placed in the regions which sell for the highest market price."[52]

Curtis linked the basic anatomy (fig. 3.8) of cattle with the slaughter animal's body hanging on the rail (fig. 3.9). He labeled the cuts to assist the stu-

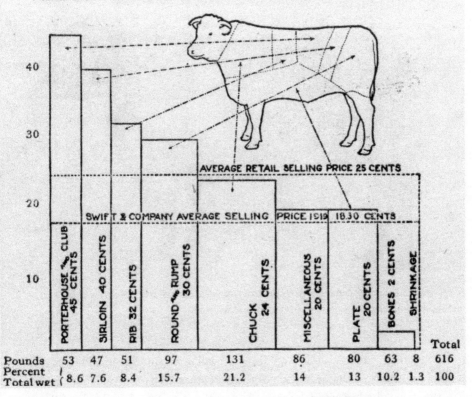

Variation in Retail Prices for Different Cuts of Beef

	PORTERHOUSE & CLUB 45 CENTS	SIRLOIN 40 CENTS	RIB 32 CENTS	ROUND & RUMP 30 CENTS	CHUCK 24 CENTS	MISCELLANEOUS 20 CENTS	PLATE 20 CENTS	BONES 2 CENTS	SHRINKAGE	Total
Pounds	53	47	51	97	131	86	80	63	8	616
Percent Total wgt	8.6	7.6	8.4	15.7	21.2	14	13	10.2	1.3	100

AVERAGE RETAIL SELLING PRICE 25 CENTS

SWIFT & COMPANY AVERAGE SELLING PRICE 1919 18.30 CENTS

Figure 3.7. Retail value of meat cuts in the beef animal. *Source*: Plumb, *A Study of Farm Animals.*

dent and breeder in making connections between the observable body parts on live animals and retail meats. Getting the breeder to understand the butcher's viewpoint was the key to getting them to adopt modern animal selection and breeding practices. Curtis argued that 56 percent of the marketable meat from the modern steer came from the rump, rib, and loin and that those cuts possessed the highest retail value as well. The overall weight of the animal when butchered thus mattered less than where the weight was located: "Animals making the highest dressing percentage conform to the block or rectangle" in body type. These animals with a "low set broad, arched rib" and deep body carried, relative to their nineteenth-century predecessors, a "large amount of the weight placed in the regions which sell for the highest market price."[53]

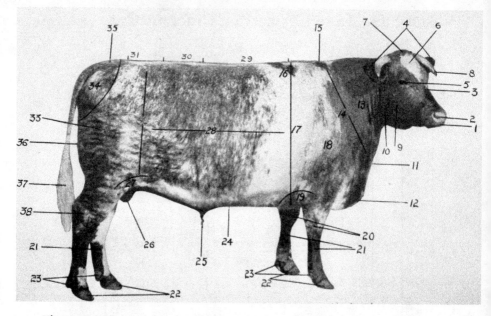

Figure 3.8. Location and names of the exterior parts of beef cattle. *Source:* Curtis, *The Fundamentals of Live Stock Judging and Selection.*

Plumb's, Curtis's, and Gay's ideas regarding mutton sheep and hogs differed little from their ideas about the beef steer. The most valuable cuts on the sheep's body were in the same areas as the steer. Plumb highlighted three places: the muscle right behind the shoulder on top of the rib (the rack), the loin, and the leg muscle. Although the International excluded most wool breeds and other reformers at the International worried little about wool production, Plumb provided standards for evaluation of fleece. According to him, a judge should spread the locks of fleece with his hand to evaluate the quality of staple and the length, consistency, and density of wool on the shoulder, back, and thigh.[54]

Gay's recommendations regarding the development of a mutton type of sheep echoed those of Plumb and Curtis for beef cattle. Butchers rarely quartered sheep when slaughtering as they did with cattle; instead they normally halved the sheep horizontally between the saddle and the rack. The back half was the saddle, which included the loin, hip, and rump. Two of the most valuable cuts of meat came from the saddle: the leg of lamb and the loin chops. The other half, or bottom half when on the rail, included the short rack and

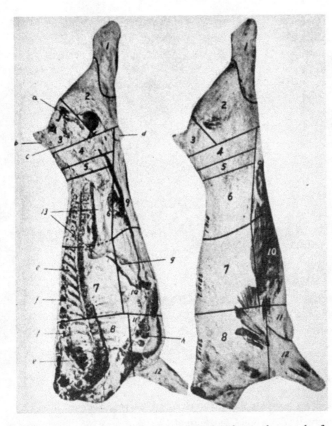

Figure 3.9. Beef carcass cuts. *Source:* Curtis, *The Fundamentals of Live Stock Judging and Selection.*

the breast. The short rack, or the rib chops, was worth as much as three-quarters of the value of the whole rack.[55]

Gay insisted that livestock evaluation should reflect the retail value of meat cuts in mutton sheep. The judge at the International, whether in the student competition or in the show ring, needed to prioritize the volume and width of the sheep from the top of the shoulder down the back to the tail set. The judges had to "handle" the rack and the loin—grab these parts with their hands—to determine their merit relative to the other animals in the class (fig. 3.10). After handling and measuring the top, judges should conclude their close inspection by handling the leg muscle. Showmen groomed the fleeces for the show ring, and the wool limited the judges' visual appraisal, requiring

Figure 3.10. Judge handling sheep to determine the width. *Source:* Curtis, *The Fundamentals of Live Stock Judging and Selection.*

them to place additional emphasis on handling and the knowledge they gained from measuring the length and width of each sheep's rack, loin, and rump in a way that was not customary for cattle or hogs.[56]

With pigs, farmers had to use different standards depending on whether they raised lard hogs or bacon pigs. Just as in the case of cattle and sheep, the marketing purpose of the pig dictated the body parts most important to selection. The scorecard for evaluating lard hogs heavily weighted the development of the forequarters, the back, and the hindquarters. The best lard hog had a broad, level, and square top that carried its muscle shape down into the ham region, which ensured the animal's skeletal frame was the widest it could be (fig. 3.11). The bacon hog differed on these points. Plumb's scorecard for the bacon type levied zero points for the development of the forequarter or the hindquarter, parts that correlated little to the overall value of this type. In-

Figure 3.11. Illustration of pig body parts. *Source:* Plumb, *A Study of Farm Animals.*

stead of being compact, level topped, and wide, the bacon hog was long with an arched back and narrower skeleton.[57]

To encourage specialization, the International featured classes that pitted animals against one another to award the best meat-producing carcass. The market classes focused on castrated animals or nonbreeding livestock, like steers, spayed and martin heifers, wethers, or barrows.[58] Organizers divided the fat steer show into two sections, one of which featured purebred steers and the other of which allowed grade or crossbred steers with purebred ancestors to compete. In each division, judges selected champions, and then they picked a grand champion steer out of the representatives from the purebred and crossbred champions. Despite many nonpurebred animals excelling in all visual metrics desired by reformers, judges had a nearly universal bias in favor of the purebred winner.[59]

Chicago meatpackers introduced the carload class to urge breeders to consistently produce high-quality animals. The International challenged stockmen to bring a carload of uniform animals, which proved more difficult than exhibiting a single exceptional animal. The single-entry steer, for example, could simply be an aberration or a "freak" and thus was not necessarily indicative of consistency in the production of market animals.[60] Alongside the carload, the Union Stockyards heavily emphasized the importance of the block class. This class allowed judges, participants, and spectators to examine the quality of the carcass on the rail. Typically, the animals would be examined and placed while alive. After the judges ranked them, they were killed and judged again.[61] This allowed the students and breeders to see the product and learn how to evaluate a live animal for market production.[62]

The International was criticized by producers who argued that shows encouraged the production of excessively, even dangerously, fat animals that farmers could not afford because of the unreasonably high input costs, but most observers in the livestock world accepted the International's aspirational goal of pushing agriculture forward toward new types of animals, even if those animals in the show ring were expensive to purchase and raise. Through the standardization and dispersion of better types of animal, the exposition reshaped animal form according to market value. Showing the best animals from each farm presented a sort of futuristic display of the next iteration or model of technological development akin to the demonstration of new machines at a world's fair.[63]

These modern animals resequenced food production; they became a way to address conflict between the irregularity of farm life (because it was seasonal) and the need for regularity in industrial production. Animals were thus not just commodities but performed labor and generated positive net returns from the aggregate value of their disassembled parts, which had the effect of recasting the farm as a commercial business intended to generate a profit from off-farm exchanges. The farmer could not eat the animals; in this industrial, profit-driven sequence, livestock had to be sold for the farmer to benefit from the value generated by animals converting Corn Belt grains to harvested meat and by-products. These modern animals served as nonhuman specialists in this growing network of agricultural experts. Simply put, modern meatpacking needed and relied on animals doing their part in the industrial chain by converting field corn to meat. The demands of this industrial progression meant that producers had to reimagine animals as technology in order to rework their physical shape.

From Dreaming to Teaching
the Animal Fantasy

The novel feature of the Exposition was the entrance of the agricultural
college and its students.... It means a new agriculture. The old is
moribund; it will die in time. The new has taken root so deeply in the
young heart of American agriculture that its future is fixed. Feeble in its
beginnings, the butt of jests and jibes scarcely a decade ago, the college of
agriculture has projected a force into farm life that is working revolution.
                    A Review of the International Live Stock Exposition, 1900

In 1915, Purdue University, a perennial participant in the Chicago exposition,
held its first Little International with the aim of showing students, faculty, and
the public the successes of "modern" livestock breeding.[1] The Purdue University
band performed, and in between the showing of cattle, sheep, and swine
in the Livestock Judging Pavilion, student organizers offered spectators side-
show acts that included fainting goats, greased-pig contests, and horse-hitch
competitions. The event closed with the capstone grand parade of livestock.[2]
John H. Skinner, dean of the School of Agriculture, had sent out an open in-
vitation and carefully choreographed every minute of the event, wanting it to
capture the attention of attendees and to mimic the rigors of the Chicago show
so as to prepare the students and animals for it. But the Little International
also advanced a curriculum and pedagogy of its own. As an intermediary between the national show circuit and the ordinary farmer, Purdue connected
the public to advancements on the university's farm, discoveries at the experiment station, and methods of modern husbandry taught in the classroom.

This energy and enthusiasm were present at many other land-grant campuses as well. Collegiate agricultural clubs with names like Saddle and Sirloin, Block and Bridle, and Hoof and Horn proliferated across the United
States, drawing students into livestock improvement. All these campus clubs
instituted student activities in fitting, showing, and judging livestock in prepa-
ration for the year's International in Chicago. Chief among the clubs' responsibilities was hosting each school's Little International, and many land-grant
schools had a Little "I." Ohio State University's Saddle and Sirloin Club, for
example, began hosting a Little International in 1912, and two years later, Iowa

State College in Ames followed suit, and eventually schools from Pennsylvania State College and the University of Wisconsin-Madison to the University of California, Davis, were hosting them. These Little Internationals advertised the advances made within each school's emerging agricultural program and offered spectators the opportunity to familiarize themselves with the roster of animals.[3] For the universities, these events also served as a direct means to a specific end—success at the International.

Despite their educational and altruistic goals, university officials, with students and animals in tow, went to Chicago to beat or best the other colleges. In the exposition's first year, schools from Iowa, Indiana, Minnesota, Michigan, Wisconsin, and Ohio participated. By the end of World War I, many others joined the competitions from as far away as California. This visceral interest in beating rival universities often led schools to cast the goals of pure science aside to gain an edge in the show ring. For example, many universities bought animals from private livestock producers and showed them to advertise the school's genetic and nutritional advances, even though the university had neither bred nor birthed these animals. Nevertheless, the schools garnered the desired acclaim, and such success at the International raised the profiles of the universities and their agricultural departments.[4] Collegiate judging was one of the most prestigious events. Students worked year-round practicing and honing their skills in hopes of making the university team. Each team had a coach—a university expert and official judge. The coaches trained students by the International's standards. These animal preferences influenced generations of students, who raised livestock and judged shows throughout the United States following graduation.

The interaction between student, breeder, and expert at the International thus gave the show leverage over the direction of animal husbandry. Many called the exposition the "bacon school" because of the education of college students in the revolutionary importance of properly selecting and feeding fat animals. In this regard, the show had a pedagogical function of its own that connected the farm to the market and gave students a chance to mingle with people from different regions and industries not present in their hometowns nor on their college campuses. A journalist reminded his readers that the spectators, producers, and businessmen themselves represented a remarkable success for the International: "The millionaire . . . from the East touched elbows with the cowboy from the range" and the "city folk filled the cup overflowing." This cornucopia of actors interacted with students and engaged with public-funded schools and government agencies devoted to agricultural reform as well as to meeting the demands of the Chicago meatpackers.[5]

At the International, land-grant officials also organized agricultural associations that promoted advancements in husbandry.[6] Charles F. Curtiss, for example, who served on the board of directors for the American Breeders' Association, which promoted eugenics, and presided over the American Society of Animal Nutrition, which provided guidelines for modern feeding, oversaw the Shropshire sheep and Berkshire pig registries and worked for the National Society of Livestock Record Association.[7] His resumé reflected the broader commitment among many land-grant professors who created and presided over associations devoted to crop and livestock improvement, organizations that revolved around Chicago.

Meatpackers depended on the efforts of these university researchers, who served as official judges, superintendents, and show managers, doing the daily work that made the International possible. And they instilled the meatpackers' standards in the minds of college students in the classroom and through extracurricular activities. This partnership at the International between government-supported land-grant universities and the Union Stockyards relied on the meatpackers' facilities and money. But the professors and university researchers did the grassroots work at the International and at their home institutions.[8] This public-private partnership resulted in a feedback loop: outputs of the International became inputs in university classrooms and experiment stations, and on farms, and their products in turn became inputs in Chicago again in the form of slaughter animals for butcher and show livestock at the International.

The universities themselves also participated in animal contests. They competed in the same classes as improved farmers and wealthy businessmen who invested in stock breeding as a hobby, like the beer-brewing Pabst family and the pharmaceutical giant Eli Lilly. With public money backing them, these universities that had hordes of student workers, professional herdsman employed by the colleges, and their own grain farms and research institutes at their disposal often dominated International competitions with their animals. Ideal body types and genetics were central to success for the schools, but they also developed regimented, precise feeding standards for their animals, thereby fulfilling a central tenet of improved livestock production promulgated at the International by linking specialized livestock production to monocultural crop farming.

The sequencing of food production began with the voluminous production of corn largely unused by humans in the harvested form, which animals transmuted to meat for consumers. According to Eugene Davenport, the dean of the College of Agriculture and vice president at the University of Illinois, this transmutation represented a shift to a scientific and money-making age that occurred following the American Civil War that disregarded the needs of the soil. The low price of food yielded by agricultural practices in the Midwest and West, the equity amassed by white farmers from land that the government expropriated from Native Americans, and the wealth generated by bankers and merchants relied on soil-robbing agricultural practices. Land "that had been thousands of years in the making," Davenport noted, had been "ruined within a generation."[9]

A by-product of cheaper food and a growing urban population was an increased demand for meat, especially beef. This demand was exacerbated by eugenicists' claim that meat consumption signified ruggedness, white superiority, and power and that vegetable-based diets were for meek, inefficient, and less civilized peoples. Although these ideas were grounded in factual errors (since, for example, corn was a staple food for American colonists while many non-European communities had diets that consisted of a significant portion of meat), the intersection of this racialized understanding of human value and diets led many eugenicists to conclude that white "superiority" and the consumption of meat helped explain the political and economic "successes" of Europeans and white Americans.[10]

These cultural assumptions resulted in a grain hierarchy with wheat at the top and corn at the bottom, a hierarchy informed by the association of Mexican and African American cuisine with corn, which overlapped with preexisting racist beliefs about non-Europeans. Hoping to save wheat flour for food aid and to feed the "fittest" Americans in battle during the Great War, food administrators tried to convince consumers at home to use cornmeal, but they were met with resistance because of racialized assumptions about cornmeal.[11] Corn thus came to be seen as most suitable as feedstuffs for animals converted to human food by animal work. Getting farmers to adopt corn-based diets for livestock became a priority for the professors. At first, the International organizers simply advertised the diets of International champions like the famous Advance in 1900. He was lauded for his full-feed diet, which included oats, corn, and oil meal three times a day leading up to the show. But they found that simply publicizing animal diets like that of Advance was not enough; farmers wanted to know why the International prioritized grain, especially corn, as part of a larger husbandry effort.

Land-grant colleges therefore created displays in a large hall at the exposition that detailed the revenue and yield advantages of feeding grain to livestock, each school focusing on one aspect of this specialization. And in 1918, the International instituted a grain show that was modeled after the livestock competitions and applied the same rates-based metrics to grain production as animal husbandry experts did to livestock.[12] The grain contest focused producers and spectators on per acre productivity, gross national food output, and the marginal costs and returns of production. Training students and future breeders in animal selection—leading them to raise purebred animals with single-purpose forms—also meant maximizing feed conversion efficiency on the farm, which demanded the marriage of modern livestock production and the use of "properly" reared and formulated grains.

On a micro level, land-grant experts used the International to curate small-scale exhibitions with animals like Advance, providing very intimate connections within demonstrations on ideal animal genetics and bodies as well as feeding and care. Swine department representative Ernest T. Robbins recalled that the exhibits "gave an unparalleled opportunity for the student to gain facility in the art of scanning at a glance large numbers of porkers and recognizing their combined excellences." The exhibits, he surmised, offered college students a unique opportunity in their "quest for knowledge."[13] These exhibits fostered personal relationships among International participants and spectators, farmers in home states influenced by land-grant work, and of course the animals whose bodies and experiences were reshaped. Reformers used the animal spectacle, the allure of the fanciful show ring, the drama of winning national honors, and the singular iteration of the superlative animal to reshape not just animals but farmer ideology. The pageantry, which captured the human imagination, stimulated fantasies about the power of human beings to recreate biological forms.

At the macro level, the adoption of new routines by producers and the aggregation of animal bodies from each individual farm helped institutionalize and normalize the requisite elements of industrial agriculture: modern, specialized animals fed monocrop corn. This process of creating new norms around a fantasy was the real work of the International. Additionally, the International provided legitimacy to these new husbandry practices and also to land-grant university institutions and work in their home states. The

International-land grant alliance lent credibility to the schools, and in turn, the land-grant experts articulated the International's industrial ideologies through their pedagogy, which they not only disseminated to students but also to farmers.

Collegiate Livestock Judging

As the superintendent of the International Judging Contest from 1906 to 1938, John H. Shepperd organized and promoted classroom and extracurricular instruction of livestock evaluation at land-grant universities that followed the standards established at the International. He worked as a professor and dean at North Dakota Agricultural College, and in 1929, he became the president of the school. While the canonical works of John A. Craig, Charles S. Plumb, Carl Warren Gay, and Robert S. Curtis, as outlined in chapter 3, provided the intellectual foundation and served as the primary texts for students and breeders, Shepperd published a work in 1922 called *Livestock Judging Contests* that outlines how to train young judges for student competitions at the International.

Shepperd credited Craig, professor of animal husbandry at the University of Wisconsin, with developing the techniques and the standards for twentieth-century animal evaluation.[14] Craig developed a new model of evaluation that turned it into a formally taught skill based on consistent standards nationwide rather than a craft that relied on the judgment of local tradesmen and breeders. Shepperd's favorable opinion of Craig also demonstrated a broader collaboration among agriculturalists at land-grant universities who shared a belief in the scientific improvement of animal agriculture and who sought to impart methodological selection to a new generation of breeders by teaching livestock judging.[15] One of Craig's most important contributions to systematic evaluation was the scorecard. His "scale of points" and his explanation of it, Shepperd writes, gave university students "a practical and scientific basis" for standardization. Craig worked for ten years on this project before publishing *Judging Live Stock* in 1901. The book became an instant success—a fourth edition appeared only one year later, and a sixth edition was printed in August 1904. Nearly all the agricultural colleges in the United States and Canada offered judging courses by the turn of the century, and Craig's volume provided these schools with their only text in the field for over a decade.[16]

Land-grant graduate Henry William Vaughan became the first International student participant to write a book about animal husbandry.[17] Vaughan received training at Ohio State University in stock judging, and he was a prominent member of the school's judging team in 1907 owing to his skill in selection.[18] After finishing his master's degree in 1909, he first taught at Ohio State and then took a job as an animal husbandry professor at Iowa State College in 1913. He subsequently taught at University of Minnesota and Montana State College and also served as the editor of the *Duroc Jersey Digest*.[19]

Vaughan represented an important generational shift. Whereas Craig focused on the judge's approach and the use of standard methods, Vaughan emphasized animal types and forms. Vaughan believed that breeding programs needed to reverse engineer animals by using "terminal" animals, that is, animals intended for slaughter for human consumption, to determine what market animals and breeding stock should look like.[20] As evidenced by Vaughan's career, the International was not simply the recipient of university ideas, playing host to their students and animals for competition, but also shaped the direction of academic literature, the education of students, and the types of animals produced on public campuses.

Collegiate livestock judging at the International provided agriculturalists the opportunity to teach students about ideal animal types, and the meatpacking companies and organizers of the International in turn relied on the competitions to create a new type of judge. Following graduation, student judges dispersed around the country and took with them new understandings of animal husbandry; they bred their own animals, and they also established the standards at other expositions and disseminated the goals of modern agriculture. Students, along with their university advisors, practiced all year for the competition, as it required students to have a great deal of knowledge and experience with horses, cattle, sheep, and swine, and many teams traveled to farms around their home states and attended other judging competitions to further prepare. Each university had five students on a team, and all team members judged four classes, cattle, sheep, pigs, and horses, each of which had five animals. This format was later changed to four students per team and four animals per class. Depending on the year, the participants were allotted either twenty or thirty minutes to judge each class before they had to move on to the next. In that time, students ranked the class and prepared written reasons detailing the different observations they made and their rationale for placement.[21]

The College Livestock Association, an external umbrella organization composed of agricultural colleges, sponsored the competition for the first four years of the show. Despite creating enthusiasm by way of collegiate rivalry, the contest received serious criticism from the most notable agriculturalists in the

United States, including Craig, Plumb, and Skinner, the last two of whom served as officers of the contest as well. Many on the collegiate judging committee expressed concerns about the process officials used to score participants as well as the competitors' penchant for cheating, as observers alleged that students handled animals from each class before the time started for evaluation and that students inappropriately shared placings with their teammates to influence the outcome of the team placings at the end of the competition. The criticisms were directed at the format, not the goals of selection. Their disagreements became destructive, nearly causing the collegiate competition to collapse.[22] In 1903, the fierceness of disagreement prompted the International to strip the college men of their responsibilities in overseeing the contest and to appoint International superintendents in their place.[23]

In 1905, the International created the office of a standing superintendent to oversee the competition, a position that afforded its occupant the same level of power and prestige as that of a specie superintendent. The first in this position was William John Black, professor of animal husbandry, deputy minister of agriculture in Manitoba, and first president of the Manitoba Agricultural College. Under his direction, the contest refound its bearings. But he quit the following year, at which point Shepperd himself took over, staying on as the perennial superintendent. To ensure the credibility and legitimacy of the program, he established a rigorous protocol for contestant behavior and evaluation, requiring students to maintain a "strict military formation" throughout the day of the competition.[24]

Shepperd created several rubrics for cross-referencing scores to ensure accuracy and eliminate errors. For example, the clerks tallied student scores horizontally and the ring scores vertically (fig. 4.1). In the end, each student could score as many as one thousand points in the judging contest, and a perfect score for the team would be five thousand points.[25] Following the conclusion of the portion of the judging in which students offered reasons for their scores, the entire staff, under the direction of Shepperd, gathered to make final tabulations that often continued into the early morning hours. This team of clerks assessed, scored, and ranked each individual and team. When completed, Shepperd and his staff posted the results in the lobby of the Stock Yards Inn between one and four in the morning. Shepperd recalled that most of the coaches were unable to sleep due to excitement over the possibility of winning the prestigious event and eagerly awaiting the results. When the officials posted the results, top teams and individual contestants would immediately send telegrams to their schools and families.[26]

INSTITUTION		OHIO						
HORSES		1st ring		2nd ring		3rd ring		Total
Student	11	30	38	40		50	42	200
,,	32	42	39	44		37	30	192
,,	53	45	37	50		50	47	229
,,	74	45	39	46		50	45	225
,,	95	45	39	50		50	45	229
TOTAL		207	192	230		237	209	1075
CATTLE								
Student	11	30	29	27		45	40	171
,,	32	50	39	45		50	49	233
,,	53	37	39	45		37	40	198
,,	74	40	34	45		37	38	194
,,	95	30	29	45		45	45	194
TOTAL		187	170	207		214	212	990
SHEEP								
Student	11	50	48	50	38	50		236
,,	32	50	50	50	44	50		244
,,	53	36	38	50	44	46		214
,,	74	50	46	46	45	46		233
,,	95	50	48	50	39	50		237
TOTAL		236	230	246	210	242		1164
HOGS								
Student	11	25	35	38		50	43	191
,,	32	22	40	40		45	38	185
,,	53	35	40	35		45	45	200
,,	74	44	44	25		50	40	203
,,	95	20	36	44		25	45	170
TOTAL		146	195	182		215	211	949
GRAND TOTAL		776	787	865	210	908	632	4178

Figure 4.1. Ohio team's scores, 1921. *Source:* Shepperd, *Livestock Judging Contests.*

According to Shepperd, the International judging contest and the rigor of the competition "fastens ideas and ideals in the minds of young men" that they retained when judging future shows or raising livestock, a claim he backed up by recalling how when he met two former contestants judging at the North Dakota State Fair and asked if they remembered placing animals in Chicago, they told him they could remember the classes of animals and the reasons they gave as "though it were yesterday." Shepperd even admitted that disagreements between officials and contestants "matters little." Instead, what counted was the "impression" on the mind and judgement in animal selection left on the student—the future judge and breeder.[27]

In 1916, Shepperd analyzed the career trajectories of former collegiate judging participants to measure their broader involvement in the agricultural community. He determined that 614 college students and approximately 100 farmers' sons (noncollegiate) had participated in the contest. Of the 439 who responded to Shepperd's inquiries, 44 percent went on to teach or conduct research at agricultural intuitions in the United States and Canada, 41 percent became farmers and breeders, just over 7 percent worked as agricultural editors, secretaries of breed associations, or livestock commission men, and less than 1 percent worked in nonagricultural professions. Shepperd concluded that the judging contest provided an unparalleled service that "assures its continuance as one of the strong features of agricultural education in North America."[28]

Although Shepperd reached out to farmers' sons for his study, the students who participated in the contests also included daughters. World War I opened the door to the formal participation of women in university agricultural courses and activities.[29] In 1917, Edith Curtiss competed on the livestock judging team for Iowa State College, where her father was dean. She demonstrated a knack for evaluating animals, ranking highly at the International, receiving recognition for being the best student on her team, and placing fifth overall. She also won a medal for placing first among all students in the evaluation of Shorthorn cattle. She became the first woman to major in animal husbandry at Iowa State College, completing her degree with honors in 1918. Following her graduation, she worked for the federal government as a scientific assistant in animal husbandry. William H. Pew, head of the Department of Animal Husbandry, lauded her long record of achievement and success. He predicted that women would replace men as animal husbandry experts at universities, even though in that field "women have been shunned until rather recent."[30]

Curtiss was not the sole woman participant. In 1917 and 1918, women from three different land-grant universities enrolled in the judging contest. Shepperd acknowledged these women's accomplishments in the contest in his annual report, remarking that the contest had been "thrown wide open" and "even been made co-educational."[31] Eva Ashton of Nebraska, for example, earned special recognition from the American Shropshire Registry Association for ranking in the top five of all collegiate participants in 1918. Ashton graduated from the University of Nebraska in 1919 and soon after joined the editorial staff of the *Breeder's Gazette* in Chicago. While serving on the staff, she published an article in the *Shorthorn in America* titled "Animal Husbandry:

A Vocation for Women" that describes a bright future for women working in the field.[32]

In her article, Ashton announces to the cattle industry that "any question as to the propriety of a woman's having a vocation is obsolete, a condition which lies at the feet of the great war." Women had been active in agriculture before the war, she argues, and made many contributions to the advancement of animal husbandry even though they did not receive formal recognition by the International or many universities. Ashton mentions the shepherdesses, milkmaids, and women gleaners of centuries past that made agriculture possible. Even before the United States entered World War I, Ashton notes, many women had owned, bred, and sold elite purebred Shorthorn cattle. In 1914, for example, the Coates's Herd Book listed seventy-one women as owners of Shorthorn cattle.[33]

In using the term "vocation," Ashton sought to question the Country Life Commission narrative of the state of agriculture that characterized farming as backward and undeveloped, which others likewise did by referring to it as a "profession." This change in language and perspective coincided with the International's and land-grant universities' goal of applying contemporary science to make agriculture an efficient, modern industry. Ashton simultaneously makes the case that animal husbandry merited respect as a career that required executive, managerial, and business expertise and also that women could effectively perform those tasks. In addition to the technocratic and commercial acumen needed, women, she argues, had the talents necessary for practical or "in-the-field" animal husbandry.[34]

Ashton died from cancer in 1921, just two years after graduating from Nebraska. Prior to her death, she made a request for policy makers and educators to help overcome barriers to women's participation in agriculture.[35] Women capably performed in these occupations, whether breeding and feeding on the farm or competing in judging contests at the International, but unequal education from the youngest of ages limited vocational opportunities for women in agriculture. She argued that women urgently needed access to schooling to achieve levels of success at higher rates.[36] Notably, as she urged her contemporaries to acknowledge the benefits that women provided the community in general, she also consistently promoted the vision of the International.

University Livestock and the Grain Competition

For universities, winning at the International raised the profile of the institution, made their livestock celebrities, and drew attention to the officials and

INDIANA BOYS' AND GIRLS' BEEF CLUB AND "FYVIE KNIGHT 2nd"
Champion Aberdeen-Angus Steer, 1918

Figure 4.2. Indiana Boys' and Girl's Beef Club with Fyvie Knight 2nd, champion Aberdeen-Angus steer in 1918, Purdue University. *Source: A Review of the International Live Stock Exposition,* 1918.

practices that hailed from their agricultural department. Skinner received many speaking invitations, accolades from university trustees, and authorship requests after Purdue University had windfall successes at the 1917 and 1918 Chicago International. The highlight of those prosperous years was the winning of grand champion steer with Merry Monarch in 1917 and Fyvie Knight (see fig. 4.2) in 1918. After winning the International, Merry Monarch became a national sensation, his picture and biography appearing in countless agricultural journals, and herdsman Jack Douglas, born in Scotland, became a celebrity in his own right. The International Livestock Exposition Association awarded him a medal for being the steer's herdsman, and the American Shorthorn Breeders' Association awarded him a medal for service to the industry.[37] Following the show, the American Shorthorn Breeders' Association donated the proceeds from the sale of Merry Monarch to the Red Cross. Skinner even bragged to Indiana governor James P. Goodrich that through this donation, Purdue and Merry Monarch might help "to do something to 'lick the Kaiser.'"[38]

Fyvie Knight also became a star after winning the International. This steer represented elite genetics, a relative of another steer with the same name who had won the International in 1908. The university purchased the older Fyvie Knight from Milton Fross of Burrows, Indiana; Purdue did not breed him. Skinner argued to anyone who would listen that the younger Fyvie Knight embodied the best practices of animal husbandry and feeding demonstrated by Purdue University. At the show and even after being butchered, Fyvie Knight maintained the status of a celebrity. He weighed 1,340 pounds and sold to Wilson and Company for $3,350. They displayed Fyvie Knight's carcass at the Biltmore Hotel in New York City. They served steaks from the steer, rumor had it, to the delegation to the Paris Peace Conference at Versailles following World War I.[39]

Land-grant universities hoped to demonstrate that the type of animals produced on university farms and the feeding programs they utilized could be replicated on commercial farms in their regions.[40] Professors wanted farmers to believe that their successes did not depend on large university budgets. These breeding and feeding practices placed value on practicality and efficiency useful to the average farmer.[41] Universities often published the process of rearing livestock that won at the International, as Purdue did with Fyvie Knight, to encourage farmers to adopt the same practices.

Skinner and Douglas weaned Fyvie Knight, born in February, after nine months of nursing. Handlers fed him cracked corn, oats, and clover hay in an open lot with a dozen other calves until December 1917. Between then and June 1918, they continued with the same ration and added some corn silage in the place of clover hay and some cooked barley. In September 1918, they slightly increased his corn intake, but they never pushed him too hard. They fed him all he could eat from September until the show in December, and they added beets and linseed oil into the ration leading up to the show. Beets or beet pulp provided the steers fiber and stretched their stomachs, which gave them more "fill" or "spring of rib" from the perspective of the judge. Linseed oil produced a glossy haircoat, enhanced fat cover, and aided in muscle development. Skinner emphasized that Purdue fed show steers feed elements that could be made from grains grown in Indiana, and thus, the Purdue regimen could be easily simulated on the average farm.[42]

Just a decade before Purdue's success, a group of reform-minded professors met at Cornell University to charter the American Society of Animal

Nutrition in 1908 and then held a second meeting at the International that same year. It subsequently became the American Society of Animal Production in 1915 and then, in 1961, the American Society of Animal Science. The founders, which included Charles F. Curtiss, were directly involved with the International.[43] The organization formed to orchestrate cooperation among experiment stations and to address imperfections in nutritional knowledge. Henry P. Armsby, for example, lamented that little scientific work had been done on nutrition as it related to the specific demands of breeding, age, and condition of animal.

The association held its first annual meeting in 1909 in the Livestock Exposition Hall at the International during which experts delivered papers and offered suggestions for future research. In his address, Armsby argued that feeding a dense American population hinged on the farmer's ability to enlist livestock in converting "grain . . . not adapted for direct consumption by man" to meat. Livestock, he contended, served man's nutritional demands by transmuting the "stored up energy of the sun's rays" in the plant. Maximizing conversion rates and efficiently transferring the sun's energy to human food was the basis of modern agriculture and the role of the farm animal.[44]

This rates-based understanding of production, in which time to market also factored, defined efficiency for the reformers. Of course, national food output was a gross measure, and increasing supply was the ultimate goal. But improving yield per unit did not directly correlate to the largest carcass or the biggest animal. These animal nutritionists wanted to measure using control groups at experiment stations the relative benefit of certain levels of feed intake and grain elements with different nutritional compositions and protein levels. These researchers hoped to improve the relative value of animals and maximize their growth rates by measuring and defining optimum feed intake levels. For example, the association's Committee on Terminology of Feeding Experiments introduced the concept of coefficients of digestibility, which measured the relative value of the quantity and quality of feed rations and grain elements to determine the peak marginal return on feed inputs in relation to daily gains in livestock. Essentially, there was a qualitative difference in inputs and a feeding quantity that hindered or improved digestibility, which could be measured by analyzing leftover or wasted feedstuffs in fecal matter.[45] Simultaneously considering digestibility and weight gains, although cumbersome, was necessary to improve not only food output but farm profit.

Animal need was an important element in this nutritional campaign. Age, species, and purpose defined the feed ration that farmers provided livestock.

Sheep, for example, had vastly different needs and utilization capabilities from cattle and pigs. But also within species, age and purpose defined the grains used and caloric inputs needed to optimally feed the modern animal. Newborn lambs started out drinking milk from their mothers, but between ten and twenty days later, they were able to digest certain grains. Young lambs required more protein to boost energy levels, improve health, and increase growth rates than did yearling or aged sheep. Mature ewes did not need the constant access to grain that young and fattening lambs did. To stimulate estrus—the period in female mammals' reproductive cycle when their sexual receptivity was optimal for conception—breeders often "flushed" ewes with grain, which was not a health requirement but a biotic stimulant to encourage breeding. Mature ewes survived on grass, hay, and silage throughout the year except during late gestation and lactation. During these critical periods, the late development of the lamb and the health of the mother necessitated the incorporation of feed in their diets, but feed that contained less protein than that used for fattening lambs. These sorts of nutritional specifications based on the animal's role on the farm applied also to cattle and hogs. Mature animals' feed requirements differed from young and fattening livestock.[46]

The US Department of Agriculture and experiment stations, which were member institutions of the American Society of Animal Nutrition, used displays at the International to define and label the types of animals and feeding regimens characteristic of the modern farm. One exhibit by the Department of Agriculture was titled "Feed Your Crops," (fig. 4.3) while the University of Illinois had ones titled "Mixed Farming" and "Livestock Farming," both of which were accompanied by model farms. "Livestock Farming" showed producers the value of livestock in maintaining soil fertility, of feeding grains to livestock to improve productivity and profits, and of building modern facilities that suited animal purpose. Many land-grant universities in the Corn Belt participated in these annual exhibits; they advertised the value of cash crops and the pairing of those crops as feed with improved livestock.[47]

The university displays were rudimentary in design and in the information provided, but they upended assumptions about feeding regimens and the possibility of increasing yields. They outlined ideas designed to replace the allure of spectacle and fantasy in the show ring with ideologies. Of course, they were just the first step in the process of encouraging farmers to embrace an industrial, profit-driven agricultural model; universities coupled the displays with university curricula, research, and publications that explained how farmers should build their farms and feed their livestock. The level of knowledge

Figure 4.3. "Feed Your Crops: Beef Cattle Use Them Efficiently." *Source: A Review of the International Live Stock Exposition*, 1921.

required to be a modern farmer—which included expertise in specialized commodity production—was highlighted by professors who insisted that university support or instruction was required for not just students but existing farmers, which led to public-funded researchers and institutions directly engaging in and reshaping commercial farm practices.[48]

Professors bragged that many farmers enthusiastically demanded practical courses after being exposed to ideas at the International. In reflecting on the International's early days, Curtiss contended that Iowa State College's success in Chicago stoked farmer demand for short courses. When Iowa won grand champion steer and the student livestock judging team earned top honors in 1902, the news spread around the state and fueled farmer interest in animal husbandry curricula. Even at this early stage, the International's prestige and goals influenced conversations among farmers who did not attend the contest nor a land-grant university as regular, full-time students.[49]

At the campus in Ames, the school offered a two-week course, and three hundred farmers enrolled in the first class—three times the number expected. The farmers ranged in age from twenty to sixty and lived in student quarters. The curriculum emphasized high-efficiency animal agriculture and expanded to corn production in subsequent years. The overwhelming success of these short courses led the university to offer instruction in more remote areas of the state. These short courses in Ames and in other communities were the primary way the improved livestock movement disseminated information to the average farmer until extension agencies began their work following the passage of the Smith-Lever Act in 1914.

The United States' entry into World War I made it even more urgent to increase food production. Seed corn shortages particularly concerned public officials. The Illinois legislature responded to this problem in 1917 by creating the State Council of Defense, of which meatpackers J. Ogden Armour and J. A. Spoor were members. Governor Frank Lowden, who served as a director at the International after his tenure in political office, commissioned the organization. The council coordinated its activities with the Council of National Defense and moved quickly to address food production limitations, particularly the seed corn shortage, which it judged the "most serious in history." The Food Production and Conservation Committee worked to increase farm output by assisting Illinois farmers with the procurement of seed, implements, and labor. This committee organized a war conference of farmers and stockmen at the University of Illinois and appointed university professors to oversee soil fertility, animal production, crop health, and seed distribution. The committee warned farmers that hoarding seed or overcharging neighbors for it provided "first class aid to the Kaiser."[50]

The legislature, however, failed to appropriate funds to help this committee provide incentives or organizational assistance to farmers. In response to this problem, a group of Chicago banks, including two at the stockyards, syndicated the Seed Corn Administration and donated over $1.2 million to distribute seeds. The administration resold seed corn with "germinating power" to Illinois farmers without a profit. And with the cooperation of the Illinois Seed Corn Breeders' Association, it sponsored a "mammoth Corn Show" at the 1918 International. Judges evaluated corn ears based on performance to encourage the "breeding up" of corn varieties that improved yield and that other farmers could use as seed corn. This first corn show was limited to Illinois participants, but its success and the clear need to pair livestock with grain improvement led the International to raise funds for a permanent grain show that included all farmers. G. I. Christie, superintendent of the Agricultural Extension Service at Purdue University, organized the show. The Chicago Board of Trade provided the premiums, and upward of two thousand entries for corn, small grains, and hay were submitted for competition from farmers in twenty-nine different states and four Canadian provinces. The

Figure 4.4. Peter J. Lux, Shelbyville, Indiana. Judges awarded Lux the grand champion twenty ears corn. He became the first corn king of the International. The National Association of Corn Products Manufacturers awarded him a $250 trophy. *Source: A Review of the International Live Stock Exposition*, 1919.

show included oats and wheat, but the corn contest was the main event (fig. 4.4).[51]

Along with animal experts, commercial seed representatives and crop husbandry professors were also traveling to Chicago to meet by the end of the 1910s. This group convened at the International in 1919 in conjunction with the first Grain and Hay Show to discuss and make recommendations for the "standardization and unification of seed improvement work." These experts led by R. A. Moore, agronomist at the University of Wisconsin, formed the International Crop Improvement Association—the first interstate organization chartered to solve crop uniformity and performance problems. University agronomists as well as farmers and stockman had long been uncertain about the reliability of seeds owing to poor breeding practices, lack of information, and intended and unintended mislabeling.[52]

At this meeting C. P. Bull, agronomist at the Minnesota Agricultural Experiment Station, delivered a lecture on the formation of pedigreed or purebred seeds through which he hoped to convince researchers and commercial breeders that it was necessary to raise seeds and label them based on the varieties' ability to cross-pollinate—an instructional designation aiding farmers' agronomic decision-making. Thus, he called for a rigorous system of inspection to ensure purity in pedigreed seed. Others in attendance echoed Bull's suggestion, focusing on the need to improve varieties based on seed adaptation, performance, and productive capacity to ensure uniformity and to increase yields. For these agronomists, standardization was not only a matter of guaranteeing the quality and performance of seeds but also one of establishing consistent nomenclature. States labeled seeds differently, using terms from "registered" and "inspected" to "certified," and farmers had little information about whether these words signified analogous field and seed inspection processes. Like animal husbandry reformers, these agronomists linked language and labeling to improvement that would address organizational or structural problems in breeding and distributing seeds.

Farmers, however, were not convinced they were being sold pure seeds that would perform in predictable ways, which dampened this agenda. Therefore, these agronomists focused on creating systems of production and inspection that verified quality and guaranteed uniformity. The association also tried to stimulate interest in elite seeds among farmers by parlaying research data into propaganda and by partnering with the International. During the 1920s, the association provided judges for the Grain and Hay Show and supported the Intercollegiate Crops Judging Contest and the 4-H Crops Judging Contest at the International and helped run exhibits on new crop varieties. These late-added competitions that encouraged farmers to use high-yielding grain varieties complemented the livestock improvement movement.

Packers, professors, and the US Department of Agriculture worked together at the International to display ideal farm setups and published and distributed blueprints to help producers remake their farms. Spoor and A. G. Leonard went so far as to purchase property in Indiana to create a utopian farm. They partnered with Purdue University, whose professors and students constructed the International experimental farm. Purdue first outlined the model farm in 1919, and over the course of three years students and professors created displays titled "As It Was," "As It Is," and "As It Will Be" to show the progress. At the International, not surprisingly, the farm featured "elite" livestock, but Purdue University also invested resources in permanent structures that systematized production and normalized profit-oriented farming practices. The university workers reshaped fields so that they were rectangular, erected production-specific structures, and built permanent fences.[53] For Spoor and Leonard, these structural changes were an essential step in developing the modern farm.

The US Department of Agriculture and land-grant professors urged commercial farmers to link building design to farm output goals. Barns and feedlot layout depended on the breeding, maternity, and fattening proclivities of cattle, sheep, or hogs, and livestock purpose also shaped farms and producer behavior. Cattle geared toward the surplus production of meat, for example, had different needs from dairy cows, and so facilities had to be constructed in such a way as to be able to accommodate different chore routines, daily animal-handling requirements, and feed demands.[54] The department emphasized that no one type of barn met all the conditions of the farms. E. W. Sheets, senior animal husbandman for the Bureau of Animal Industry, and M. A. R. Kelley, barn architect for the Bureau of Public Roads, jointly published an article in 1923 that urged farmers to adopt certain designs based on climate, topography, and soil needs (fig. 4.5). Open-sided barns or barns with doors for ventilation, for example, should be exposed to the southern horizon to protect against the cold winds and winter precipitation that approached from the north and west. The dictates of climate determined the choices farmers made, and farmers had to find a site for construction that through proper drainage would keep the pens free from standing water. Constructing lots and pens to improve manure collection capabilities was essential. Building concrete pads, reservoirs, or barricades to prevent runoff and maximize manure retention helped farmers in the collection, processing, and redistribution of fecal and vegetable matter on fields, which helped maintain soil fertility.[55]

Furthermore, livestock had species-specific requirements (fig. 4.6). What separated cows and sheep from hogs was their digestive needs. Cows and sheep are ruminants with compartmentalized stomachs that only function

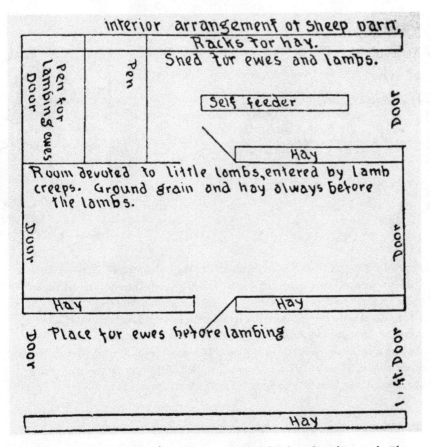

The diagram contains the following handwritten labels:

- Interior arrangement of sheep barn. Racks for hay.
- Shed for ewes and lambs.
- Pen for lambing ewes
- Door
- Pen
- Self feeder
- Door
- Hay
- Room devoted to little lambs, entered by lamb creeps. Ground grain and hay always before the lambs.
- Door
- Door
- Hay
- Hay
- Door
- Place for ewes before lambing
- Door
- Door. Fig. 1.
- Hay

Figure 4.5. Sheep barn for lambing. This barn divided the sheep based on needs. The bottom pen, which had access to an outdoor lot, was designed for gestating ewes. The middle pen was for nursing lambs only. The mothers stayed in the top pen and a creep gate that allowed only small lambs through separated the two pens. This allowed the lambs to eat grain at rates suitable to their age and growing demands without their having to compete with mature ewes. The top left pens were jugs, which had specifications for ewes during birth—small, tight, and warm. The small size helped keep the lamb and mother together, which encouraged bonding and nursing at that young age. The hay racks for sheep of all ages had feed bunks underneath. *Source:* Stewart, *The Domestic Sheep*, 1900.

properly with the inclusion of grass, hay, or silage in their diets; roughages aid the digestion of grain. But hogs have a single-compartment stomach and perform well without the inclusion of roughages in their diets. As a matter of fact, hogs can live off the leftover corn bits in cattle manure; some farmers at

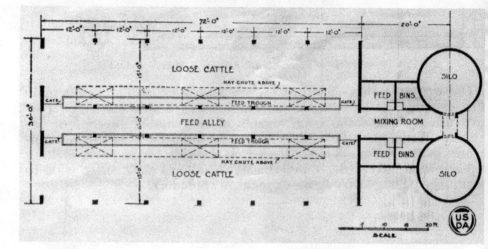

Figure 4.6. Floorplan for a cattle-fattening barn. The walls pictured on the top, bottom, and left were kept open to properly ventilate the barn and allow the feeder cattle access to the lot. The two loose cattle pens on either side of the middle alley allowed the feeder to access the elongated feed bunks from the middle of the barn, which connected to the hopper-bottomed feed bins, the feed mixing room, and the silos. Above the alley were access points to the hay mow. The length of the barn was variable based on the number of feeder cattle. Each animal needed thirty inches at the bunk. Source: Sheets and Kelley, "Beef-Cattle Barns."

the time even included a few hogs with a herd of cattle to make use of wasted corn.[56]

Design that took into account the purpose of the animals being raised initiated a transition to high-density feeding.[57] The feedlot regime that began to take shape during this period ultimately characterized cattle production in the latter half of the twentieth century, which emphasized dry-lot fattening operations with dirt, gravel, or paved yards. For instance, the Illinois Experiment Station modeled and advertised the benefits of a paved lot in connection with market value. Animals that laid, walked, and lived in mud and muck were monetarily docked at the market. Clean-hided, healthy-appearing steers garnered a premium at Chicago markets as the experiment station confirmed through its own trials as early as 1904.[58]

The relationship between land-grant institutions and private industry that developed in the first years of the International grew with each passing decade. By the end of the twentieth century, their financial and research interests were interwoven. Seed corn companies financed professor research, funded university construction projects, and sponsored collegiate sports. The protein industry too underwrote meat science work, and in turn, the interests of university research and curricula justified and reified the types of farming practices, animal rearing and disassembly, and food distribution that was standardized at the International and that benefited the meatpackers and agribusiness more broadly. In effect, the International collapsed categorical differences between the "pure" academic research of land-grant institutions and the applied research of commercial pursuits.[59] As the public university-private industry bond strengthened, the original objectives of the university professors were undermined. In particular, the goals of balanced farming and mixed husbandry gave way to the industrial prerogative. The agricultural reforms subsidized by private industry, as represented by the meatpackers at the International and propagated by land-grant researchers normalized monocultural production and high-density animal rearing that necessitated the off-the-farm sale and disassembly of the animal. In this model, specialization in crop production and the modernization of animal husbandry were necessarily linked. The organizers of the livestock improvement movement initiated structural changes to crop and animal husbandry never intending the twentieth-century farm to be a pastoral, subsistence-based homestead—instead, reformers hoped to make the farm a commercial firm.

CHAPTER 5

New Animals, New Problems

During its twenty years the International has . . . encouraged the breeding of pure-bred live stock; it has improved the quality of such stock; it has popularized the judging of livestock, and through its efforts many young men have been induced to enter this useful field. . . . In these troublous times many would despair of the future of the Republic, if it were not for the millions of self-reliant, patriotic American farmers.

A Review of the International Live Stock Exposition, 1919

The International took aim at the roughness, ponderousness, and old age of nineteenth-century livestock and redirected breeder attention toward quality. This approach that focused on animal value differentiated nineteenth-century husbandry from the practices that reformers considered "modern." Defining value, however, proved more difficult than simply recognizing the weaknesses in animal agriculture. The International intervened to frame the conversation and construct the standards for the improved livestock movement. The show provided examples of modern animals, and through a large group of supporters and surrogates, including the agricultural press and land-grant universities, the International served as a hub in a feedback loop that incentivized the scientific breeding and feeding methods used to propagate and raise "superior" animals.[1]

In 1919, professors, packers, and agricultural journalists celebrated the twentieth anniversary of the International. Publications circulated among farmers and ranchers praising the International for having succeeded in making scrub livestock unpopular and unfashionable. James Poole and other industry insiders extolled the displacement of unknown or mixed genetics in favor of the "superior" blood of British livestock.[2] The International, Poole maintained, had done more than any other institution in forging an agriculture less susceptible to cycles of instability or depression; its founders had successfully "designed an institution of permanent character for the improvement of the live stock of the new world."[3]

Even though the agricultural press took this opportunity to applaud the International fairly unanimously, questions soon emerged about the unexpected consequences of livestock improvement. Poole himself, for example, modified his stance. To be sure, he remained committed to the goals of the show, but he began to worry about the gulf between the body types most appropriate for the meat industry and the type of animals increasingly selected in the show ring. Beginning in the 1920s, the show-ring ideal—new-aged, moderate-framed livestock—resulted in even smaller cattle, engendering the baby-beef era. In effect, the International had failed to standardize animal form in the ring and instead initiated a long cycle of insular physiological transformations. Over the next four decades, judges kept selecting smaller and smaller animals, and breeders continued raising them to win exposition accolades. By the late 1940s, champion steers stood waist high by their handlers. These compact cattle also spawned a "dwarfism" outbreak. Unreasonably small livestock had physiological deformities, did not reproduce at high rates, and yielded uncompetitive carcass sizes.

The show industry responded to the problems caused by extreme compactness, and by the late 1960s, judges began to select for structurally longer and taller animals; they even exhibited a willingness to crown a crossbred champion on what they called "performance" grounds. But this show-ring shift toward bigger animals was just another fad, the "frame race," that culminated in extremely tall, long, and often narrow cattle, sheep, and pigs by the 1980s and 1990s. The oscillation in animal form in the show ring over the decades had the effect of separating specialization from standardization. The International succeeded in normalizing industrial ideologies and the corresponding purebred livestock specialized in meat production.[4] Farmers indeed adopted purebred animals, marketed younger livestock, and changed production methods, which led to broader specialization in agriculture. However, the International failed to create a consistent standard in the show ring, and ultimately lost its place as arbiter of livestock form.

By the 1960s many commercial breeders had become frustrated over the inadequacies of champion show animals and the high costs associated of raising them. Maturing agricultural institutions, like land-grant universities, experiment stations, and extension agencies, as well as agribusiness began a transition to performance data—actual and potential carcass traits, rate of gain, and maternal performance—to determine animal merit. The result of the analysis of this data was that commercial farmers began to breed moderate-sized livestock on specialized farms while the International showmen continued to raise rare, expensive-to-produce animals that edged toward extreme and impractical sizes, resulting in serious health and economic consequences.

Prohibitive Costs and Impractical Animals

Poole's criticism of the International just two short years after he had un-
abashedly praised its success created ripples in the industry. Specifically,
Poole had condemned the International for inflating the monetary value of
"improved" animals, which created cost barriers for the average farmer. This
was a "rich man's game," Poole alleged: a small circle of breeders merely
swapped or sold elite animals among each other with no regard for typical
farmers or their farm conditions. The only access to this fraternity of breeders was "an overwhelming desire for high-priced livestock, and a loose pocket-
book." For average farmers, paying interest on mortgages and affording farm
inputs while making a living wage quelled "dreams of avarice."[5] Indeed, the
show ring drove improved animal prices higher and stoked owner ego, under-
mining the show ring's practical usefulness for farmers. The very process of
selection—prioritizing animals with superlative qualities—created a secondary market for show livestock unattached to market fluctuations and commercial value.

To justify this hobby, breeders of extreme animals cloaked self-interest and
pride in public service narratives, regarding their participation at the International as demonstrating patriotism. World War I helped their cause, as the
US Department of Agriculture declared the International the "food training
camp of the nation."[6] Herbert Hoover, who ran the US Food Administration
at the time, heralded the International's central role in this effort. For Hoover,
food was no less important than ammunition in conducting the war. Consumption during the war greatly outpaced supply, and thus the focus on gen-
erating more food and by-products per animal at the International appealed
to Hoover. Better-quality, early-maturing animals ensured maximum yields at
a reduced marginal cost, presenting a solution to the broader food shortage
problem. The production of meat directly contributed to the war effort, the
International crowed, and the organizers of the show validated producers following the armistice by congratulating their purebred animals for aiding in
victory. These wartime accolades reinforced producers' and reformers' belief
that improved livestock drove agricultural advancement and that agriculture
foundationally supported and propelled American prosperity and security.[7]

Poole conceded that because of the International, even the common
breeder no longer doubted the value of purebred animals. The International
had convinced a generation of stockmen that "mongrel" genetics compromised farm revenue and national agricultural yield. But Poole nevertheless

concluded that even though farmers accepted the central tenets of modern animal husbandry, their ability to acquire purebred animals was limited, which meant that reform goals were not achieved. The reason was self-evident: inflated prices. At the time Poole wrote his critique, cattle prices on the market had plummeted. The agricultural community experienced hard economic conditions in the early 1920s. In that context, Poole's concerns were understandable: "better sire campaigns" and "purebred sermons" did not put money in the pocket of normal farmers, especially since they could not touch those price levels "with a ten-foot pole."[8]

Breed associations and the agricultural press pushed back against Poole's critique by first questioning his character. The Aberdeen-Angus breeders alleged that Poole only published his criticism in the *Producer* because western producers resented the International's denigration of range livestock and husbandry practices in the West and its simultaneous pushing of overpriced "superior" animals.[9] Journals and breed associations also argued that Poole overlooked valuable improvements in commercial operations that followed from "breeding up" when purebred sires were mated with inferior cows; the crossbred offspring increased the net value of livestock herds on the range, carcass yield, and producer revenue (fig. 5.1).[10]

Purebred Hereford cattle, for example, had red bodies with distinct white faces and sometimes white markings on their feet and tails. When crossed with other breeds or "mongrels," Hereford bulls stamped their offspring with white faces. White-face calves with black, brown, and red bodies filled the range in the twentieth century.[11] Range producers adopted many of the tenets of progressive husbandry, and the use of purebred bulls shaped the genetic makeup and aesthetic look of their cattle, so much so that areas around Kansas City were unofficially known as "the Herefordshire of America." In 1919, 75 percent of cattle that passed through the Kansas City livestock facilities had the Hereford's characteristic face.[12]

Representatives of the Aberdeen-Angus Breeders' Association also countered that the International deserved recognition for encouraging farmers to raise younger animals with bodies better suited for feedlot production. The International's power in changing farmer behavior became its enduring legacy in the improvement movement. For example, to increase efficiency in production, packers and professors encouraged farmers to feed grain in more dense animal populations, which led to a greater presence of feedlots and an increase in hornless cattle, which were preferred because cattle with large horns performed poorly in tight quarters and caused problems on the farm,

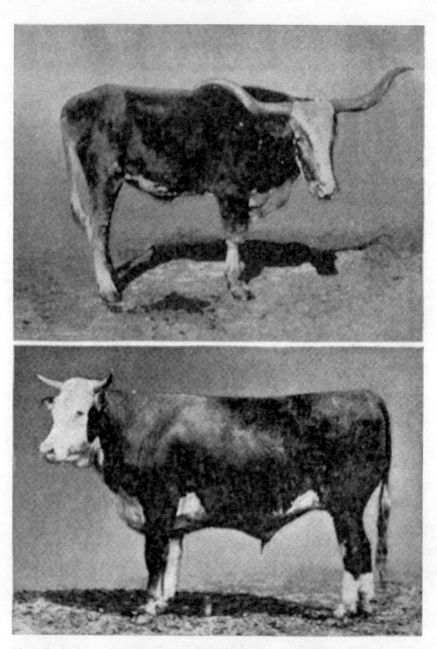

Figure 5.1. A photograph published by the USDA in 1919 to announce the near extinction of the Texas Longhorn owing to the ascendency of superior livestock—Herefords. *Source:* Sheets, "Our Beef Supply."

during transport, and at the stockyards. The connection between improving efficiency and eliminating dangerously large horns circled back to the importance of British purebreds. Most British cattle breeds either had smaller horns or no horns.

All in all, better-breeding campaigns succeeded in increasing the number of purebred cattle in the United States. Between 1908 and 1914, purebred cattle numbers in United States fluctuated between sixty thousand and ninety-five thousand registered each year. By 1957, the number of purebred beef cattle registered annually had grown to 860,000. It was the same for hogs and sheep. In the sheep industry, producers used purebred genetics to increase wool and meat yields. Fleece weight, staple length, and clean wool output all increased from 1900 to 1950. This upward trend in fleece production mirrored the commercial emphasis on meat output. The overall numbers of lambs per one hundred ewes dramatically increased as well, which allowed farmers to raise more meat with the same number of ewes.[13] In the swine industry, better breeding changed the finish age of market hogs. Over the first half of the twentieth century, pigs reached slaughter weight two months younger and converted feed at better rates. Feed required for a market hog to gain one hundred pounds decreased from one hundred to eighty pounds. Changing farmers' practices demonstrated the International's contribution to the national livestock improvement campaign. Even though the range did not fill up with show animals from the International, as the *Aberdeen-Angus Journal* reminded readers, producers changed their methods of selection, breeding, and culling based on the standards established in Chicago.[14]

Specialization

Specialization in one area of agriculture—the animal body—influenced all aspects of farm structure. The International pushed for a high-yielding, single-purpose livestock, which affected every level of production and the tasks and commodity focus of the farmer. Not only did farmers become reliant on off-the-farm inputs and nonfarm agricultural experts but producers and their animals also became specialists within this broader agricultural sequence.[15]

Packers pushed specialization because they hoped it would provide a consistent supply of quality livestock to the Union Stockyards, but it was not necessarily conducive to balanced or permanent farming. Stock feeders bought animals and feed, and they generated revenue solely from off-the-farm exchanges or sales and had little ability to return the manure and waste of animals to the soil where the grains grew, and even when manure that collected in large quantities on specialized livestock farms was able to be safely distributed at low rates to the grain farms, leaks and spillage of fecal matter from manure reservoirs presented an environmental hazard to farms and local communities. In addition, the revenue demands of single-commodity production skewed the principles of permanent agriculture; by 1939, 80 percent of midwestern farms produced corn, most of which went to livestock, whereas previously they had raised fruits, vegetables, and animals.[16] Even when animal production did not yield a profit, the fertility gains of balanced farming had negated financial losses.[17]

While the professors regarded the use of livestock on every farm as the way to ensure the continued productivity and fertility of the soil, the packers saw commercial fertilizer as the solution, which was not surprising, as they owned commercial fertilizer companies. Commercial fertilizers provided important elements that were mined or extracted and transmuted into synthetic compounds, allowing producers to maintain and increase farm yields by invigorating crop growth. However, many of these fertilizers only stimulated crops and did little to build the soil structure.[18] Despite this, yield gains that resulted from the application of commercial fertilizers drove farmer reliance on them and ultimately undermined the professors' balanced-farming goals.[19] The limitations of land expropriation, range expansion, and extensive farming that caused soil exhaustion in the nineteenth century carried over to the modern farm, which relied on extracted materials to make viable new versions of soil-robbing husbandry.

As a geographical middle ground, the Great Plains became a contested space for differing production regimes. Would it adopt range-like husbandry practices, or would it be subsumed by the growing influence of Corn Belt agriculture and the feedlot system of production? Geographer Terry G. Jordan has called this a "cultural contest" between Texas and the Midwest, a contest that the Corn Belt eventually won. Producers provided feed to herds in the winter and marketed animals year round. The docility of animals, the limited need for equestrians and equestrian skills, and the dependence on permanent livestock facilities—like barns, silos, and fences—differentiated this regionally based agricultural system, which manifested in cattle fattening districts (fig. 5.2).[20]

Nothing was more indicative of this transformation than farmers' growing dependence on commercial inputs, including feed. The feed manuals published at the end of the nineteenth century, like Elliot Stewart's *Feeding Animals*, had little to say about manufactured or mixed feeds, but just three

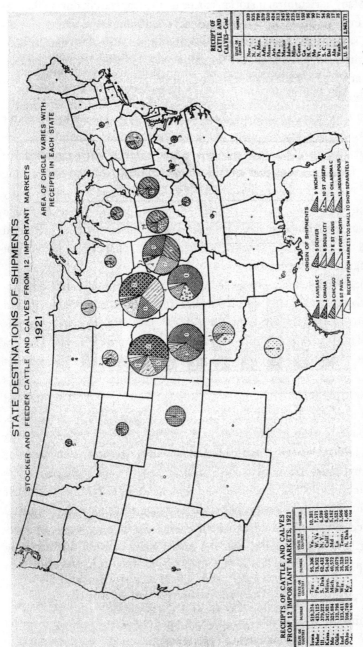

Figure 5.2. USDA map that depicts the percentage of stocker and feeder cattle bought by state. After they were purchased, these cattle returned to the farm for fattening. Overwhelmingly, the Corn Belt served as the fattening district, especially Iowa, Nebraska, Illinois, Kansas, and Missouri. *Source:* Sheets, "Our Beef Supply."

decades later the commercial feed industry was grossing $400 million in revenue. Many agricultural historians focus on mechanization during this period, but in 1929 the feed industry outpaced agricultural machine and attachment sales by $122 million.[21] Indeed, the International turned animals into machines themselves that efficiently converted feed to meat and that yielded more food per acre and increased aggregate food output, and the professors who created, officiated, and endorsed the International used the exposition to advertise the need for improved nutrition and to develop a better understanding among producers of the biological needs of animals.[22] The push toward the formulation of specialized feed—feed produced off the farm with other farmer's grain—further undermined the goals of balanced farming and mixed husbandry. The expert knowledge needed to produce scientifically manufactured feed and fertilizers and to design and build the newest farm machines and attachments forced specialization on the farm, which increased the capital demands of farm production and farmer reliance on a network of nonfarm agricultural specialists.

Livestock producers invested in concrete floors for animals and built permanent structures for a specified purpose, including for animal birth and feeding. One progressive farmer in Illinois, S. W. Larmore, reported to a *Prairie Farmer* journalist in 1918 that he bought feed from local tenant farmers and that he had built a series of structures for his hogs. He also bragged that he had enough concrete on his farm to pave an entire Illinois town. Larmore's investment in permanent structures exemplified the type of farm normalized by International promoters. While he fed hogs in tight pens on concrete floors, many other farmers used slatted floors. In addition to feeding facilities, hog producers built special buildings called farrowing houses to birth piglets. For his farrowing house, Larmore had a series of eight-foot pens under one roof.[23]

Such function-oriented facilities with specific designs typified Corn Belt meat production.[24] Like feeder hogs, feeder cattle and sheep also required constant access to feed and storage facilities for grain and roughages. Smaller-framed cattle, in particular, hastened the development of feedlot. The feeding regimen for so-called baby beeves provided a stark contrast between modern husbandry and the range. Baby-beef cattle had small, compact bodies with short legs. Not only were these bodies incapable of thriving in the grass-fed, range regime but also the amount of travel required on the range prevented baby beeves from optimizing the calories they consumed.

The packers and professors who organized the International were pivotal in initiating this trend and normalizing high-density production systems that over time grew in scale and dependence on commercial feeds. But high-density feeding, whether on the feedlot or in enclosed structures, caused unintended problems. Dependence on grain calories to fatten livestock led to overeating and acidosis, or grain overload, in ruminant animals, which required the utilization of roughages for digestion and to counter grain imbalances. Further, even on closely monitored and micromanaged feedlots, larger and stronger animals pushed weaker ones aside while eating, preventing the feeder from dictating the amount of grain each animal consumed. In these situations, the better-performing animals, because of their strength and higher rate of grain consumption, became more susceptible to disease. Pathologists and veterinarians advised feeders to separate cattle and sheep based on age and size so that the animals competed fairly, which made it easier to regulate feedings.

The biological responses of cattle and sheep to the feedlot system pushed producers to provide additional support to guarantee the survival of the animals and reap the economic benefits of feedlot production. Feeders inoculated or vaccinated sheep, for example, as they arrived at the feedlot to cut down on the prevalence of enterotoxemia—which is overeating—and coccidiosis, a disease caused by parasites of the *Eimeria* genus, and during a suspected outbreak, the feeders would force their animals to eat sulfur, which dramatically reduced the number of lambs that died after experiencing symptoms.[25] To deal with parasitism, a problem caused by the fact the feeder animals lived on top of their own fecal matter, which facilitated the transmission of internal parasites that prevented them from utilizing the nutrients available in the grain, farmers invested more money in feedlot infrastructure, altered management practices, and adopted a parasite treatment program.[26]

Fads in Animal Form

International judges' preferences for superlative animals deviated from the ideal of a standard body type and distorted the market. By definition, superlative animals were rare and thus too expensive for commercial producers, and this drove a wedge between the show ring and the commercial farm, giving rise to two livestock industries and biological types: show animals and commercial livestock.

The baby-beef era perfectly captures the mixed consequences and cumulative effects of the International and the show ring. Breeders and judges in Chicago used a specific set of parameters to identify and reproduce "baby beeves," or the "pony types." Baby-beef steers typically ranged from twelve to twenty-four months old, although some farmers more narrowly defined the upper age as being fifteen months.[27] These steers weighed eight hundred to twelve hundred pounds and were either choice or prime in meat quality. Slaughtered baby beeves furnished thick, light, and flavorful steaks, which the twentieth-century consumer preferred.[28] Despite the higher overall weight of a two- or three-year-old carcass, the yearling animal yielded a higher percentage of quality meat, 25 to 50 percent more meat than their older counterparts. Prioritizing yield over gross weight in order to address the need for more efficient animals and better-quality meat distinguished the baby-beef era.[29]

Boys' and girls' clubs organized around baby-beef goals, and the International hosted their baby-beef projects starting in 1916. Whether in the youth competitions or in the show ring, these types of animals were normalized by International judges. They overwhelmingly selected yearling cattle, which effectively ostracized older steers.[30] In the carload class, for example, every champion group after 1908 was between twelve and twenty-four months in age. From 1900 through 1908, the average weight of the carload class winners was 1,457 pounds. Over the next nine shows, the average weight dropped to 1,138 pounds.[31] At the International's sale and on the commercial market, animals weighing between nine hundred and twelve hundred pounds received a premium over heavier livestock—an inversion of market preferences from decades before.[32]

The International fell victim to fads, failing to distinguish between functional animals capable of mass reproduction and livestock manufactured to appease fanciful tastes. Because judges had to create, at least in their minds, a taxonomy of important qualities and select animals that possessed the best part or parts that were highest on the list of important qualities, they looked for the biggest backs, deepest rumps, and increasingly the shortest and most compact bodies, which in turn pushed breeders to extremes. Perhaps competing animals possessed better color, breed character, heads, or constitution, but because the judges' main concerns related to market value, they started with body parts that correlated with carcass and worked down their lists of priorities.[33]

Emphasizing extremes in animal evaluation usurped standardization by continually moving or resetting the mean or average in body type, as judges would select animals that demonstrated extremes based on the new norm. The extremes of the show ring inflated the prices of breeding stock, and it was these high costs that critics lamented and the commercial producer resented. This gap between commercial needs and show-ring priorities resulted in irrational aesthetic preferences as well. In the US Department of Agriculture's 1936 yearbook, W. H. Black of the Bureau of Animal Industry argues that aesthetic qualities, traits that help breeders identify breed, distracted the industry from moving toward breeding more economically productive cattle, although he commends the improved livestock movement for identifying breeds that specialized in a purpose. In his view, aesthetic traits were useful for identifying meat-producing breeds and selecting mating pairs that improved output, but head, nose, hoof, hair, wool, hide, and color possessed no tangible market value. The value attributed to aesthetics in the show ring, however, distorted these animals' value to the average farmers.[34]

This gulf between the show ring and the needs of the average farm became overwhelmingly apparent when Arthur Weber selected Old Gold for the Grand Champion Steer award at the International in 1948, effectively ushering in a new era of ultrasmall, extrafat cattle. Weber served as the head of the Department of Animal Husbandry at Kansas State University starting in 1944 and later became the dean of agriculture and director of the Kansas Agricultural Experiment Station. Weber judged the steer show at the International for eleven consecutive years, and as a result, his preferences for ever more compact animals directly shaped the trajectory of animal fads at the International. The fatter bodies were of commercial use during WWII, aiding in the production of tallow needed for explosives. However, with every new fad came a new set of problems.[35]

By midcentury, International steers had become so small and compact that agriculturalists developed new words to describe the animals. "Compact" or "Comprest" steers emerged in the late 1940s and 1950s; reformers and judges also referred to them as "belt buckle cattle," because they wanted them to be no taller than waist high next to their human handlers. These words and terms developed into categories of type or form that influenced judges' selections, and the extreme shortness that the show encouraged reduced the commercial productivity of the animals. Conventional Shorthorns, for example, were able to convert digested dry matter into more commercial products than Compact Shorthorns. Even though judges and breeders associated smaller types with greater productivity, the Compacts and Comprests did not provide any carcass advantages over conventional purebreds in calorie efficiency, carcass yield, or product value. The selection of ultrashort cattle incidentally caused a reduction in overall body mass and, thus, a diminished ability to meet market demands.[36]

The original plan of the International to moderate size and improve carcass yield of livestock by discouraging the reproduction of the large, thin scrubs with unknown genetics of the nineteenth-century range initially worked. In the first two decades of the twentieth century, agriculturalists reported commercial improvements in livestock. But the show ring failed to reward moderation or balance, and the short-term usefulness of smaller animals ended when livestock pushed passed practicality to unreasonable shortness and fatness. For example, "breeding up" campaigns required farmers to use "improved" males, but the bulls' short stature prevented them from successfully mating with even average-sized cows. To assist reproduction, producers dug holes in the ground in which the larger female stood to lower her height and allow the "improved," extremely short bull to mate with her. Excess fat limited vigor, reproductive health, and longevity. Of course, in the show ring, the additional fat carried by these Compact cattle gave the appearance of a greater degree of finish, which was visually appealing, yet deceptive. The fat created the illusion that the animal carried more consumable meat, a desirable feature for judges. But the short strides and overweight bodies of these cattle inhibited their on-the-farm performance.[37]

This pattern repeated itself with hogs. The externalities of the show ring resulted in overly fat pigs or "improved boars" with vastly different body types from average sows, which required reformers to reevaluate the feasibility or the efficacy of natural breeding. Breeding crates became the solution to this problem, facilitating mating between sows with leaner body types and the fat, square-made boars from the show ring. The crates did not force reproduction; instead, they protected the well-being of female and male pigs, preventing the boars from chasing nonovulating females that were not willing to mate, which preserved the health of expensive boars, sows, and gilts. Boars only had access to sows and gilts once they reached peak fertility. The farmer identified an ovulating female by observing physiological indications of estrus—like erect ears, a willingness to stand or mate—and then the farmers moved the females to the crates to aid in reproduction.[38]

The crates varied in design but were generally box-shaped with an opening in the back connected to a small, narrow chute that dead-ended with two ledges on either side of the sow (see fig. 5.3). Some crates had irons that prevented the sow from falling during mating while supporting the boar; some simply allowed the boar to guide himself with his front feet by using the ledges. The crates were especially necessary in situations where the boar was much heavier than the sow. Crates made it possible for farmers to manage and control breeding, which not only allowed them to determine the date of birth but also helped them ensure that their males and females were

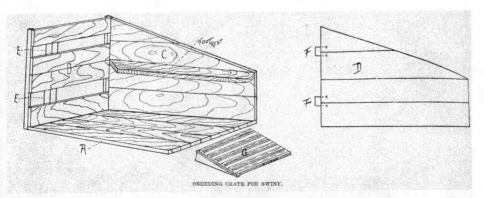

BREEDING CRATE FOR SWINE.

Figure 5.3. Basic breeding crate. *Source: Breeder's Gazette* 53, no. 4 (1908): 180.

in the best condition while mating, which increased conception rates and litter size.[39]

In fostering a more productive environment for propagation, crates protected investments made by producers in "improved" animals by reducing the risk of injury or harm to sows and boars and by allowing boars to mate without exerting much energy, which meant these expensive animals could breed more sows without tiring. Because the crates enabled boars with limited on-the-farm functionality to reproduce at high rates, they created a cycle of breeding more and more animals with serious physiological flaws. This succession undermined animal productivity and also encouraged farmers to propagate and keep breeding stock that endured painful structural problems.

The show ring also spawned an outbreak of dwarfism in the cattle industry in the 1950s that caused hysteria among farmers. This problem had begun twenty years earlier when judges began selecting for "duck-legged" steers that had ultrathick bodies and excessively short legs; at this time, judges regarded these animals as the most extreme examples of the idealized body. But as breeders oriented their matings based on this standard, the often lethal "dwarf" recessive gene emerged in purebred animals, especially prominent in Herefords and Aberdeen-Angus cattle. Not only were the smaller carcasses a drawback at the market, but young calves had bowed front legs, short, flat faces with protruding foreheads, undershot jaws, and pot bellies and experienced incoordination, nervousness, and breathing difficulties that earned them the nickname "snorters," compromising their well-being and life expectancy. Snorter Shadow Isle Black Jestress 2 won the Grand Champion Angus Female in 1953 (fig. 5.4).[40]

Figure 5.4. Shadow Isle Black Jestress 2, grand champion Angus female at the International Livestock Exposition in 1953. *Source:* American Angus Association.

Robert Hough, former collegiate livestock judging coach, cattle historian, and animal scientist, argues that these animals were "far too small for commercial use" and that the belt buckle era came to an end because commercial producers rejected "the cattle, their clear lack of utility, and the curse of dwarfism in some of the most popular herds." Indeed, the dwarfism that injured the purebred industry was, according to him, "Mother Nature rebell[ing] against these compressed cattle."[41] In 1969, Don Good, chair of the Animal Science Department at Kansas State University inaugurated a new era in cattle selection with the crowning of a crossbred steer as the Grand Champion at the International. The steer, Conoco, was an Angus and Charolais crossbred (fig. 5.5). In selecting Conoco, Good showed a willingness to break the purebred tradition, which also highlighted a turn away from the compact types of the 1950s toward types assessed according to "performance."[42]

Figure 5.5. Conoco, grand champion steer at the International Livestock Exposition in 1969. *Source:* Robert Hough.

The International ended in 1975, but in state and national shows throughout the country, judges began prioritizing growth, height, and leanness with better yielding carcasses. This new fad, like the previous ones, turned to excess in the 1980s. "Single trait selection" for larger frames, Hough recalls, led to cattle "frame[s] and size . . . pass[ing] the point of common sense" (fig. 5.6). Because high calf weights and long legs, specifically the cannon bone, indicated that a calf would be tall, producers selected for these traits, but the consequences included that it was difficult for cows to birth calves with these frames and that the animals were impractical on the commercial farm. At "the zenith of the frame race," Hough concludes, these "severe structural problems were accepted because straightening out the joints . . . made the animal taller."[43]

Sheep and hogs went through their own timelines of oscillating size following the compact-animal fads of the 1950s. Sheep also moved to large

Figure 5.6. BB Contender, grand champion full-blood bull at the North American Livestock Exposition in 1982. *Source:* Bert Moore.

frames, a trend that continued into the 2000s. These large frames resulted in structural problems, ewe and lamb mortality, and high costs for producers. In the 1980s and 1990s, hog breeders reconfigured pigs to have long, lean bodies, moving away from short, broad, fat show pigs. But because of the limits to functionality of these leaner pigs, breeders again changed course in the new millennium. The selection of "superlative" cattle, sheep, and pigs and the problems associated with animals of extreme types created a cycle of changing animal forms divorced from commercial utility and standardization. Livestock evaluation at the International helped create this gulf between the standards meant to advance commercial agriculture and the individual animals selected as champions. International animal evaluation pushed livestock to extreme and even dangerous forms, which made them rare in quantity and stimulated high prices. The physiological problems associated with small, compact as well as tall, "frame" cattle thwarted efficiency goals, resulting in cost, reproductive, and animal well-being problems.

While the animal fantasy portrayed in the International's rings was never fully actualized in American pastures, the exposition directly shaped farm structures, animal breeding, and food production around industrial ideologies. The industrialization of agriculture relied on the International's ideologies—making machines of animals—and the emerging farm factories churning out corn and soybean crops also depended on livestock to generate value from Corn Belt grains. Capital- and labor-intensive processes, dependent on nonfarm agricultural specialists and orchestrated by the International, became the predominant husbandry regime in American agriculture by the end of the twentieth century. Even though many of its champions flaunted impractical physiological shapes, the International normalized industrial ideologies and the biological homogeneity that accompanied "improved" breeding.

The Legacy of the Modern Animal

As the twentieth century unfolded, the spatial distance between animals and humans, including grain farmers, in daily life resulted in the transmutation of livestock into largely invisible commodities. The emotive distance between human and animal mattered just as much as the spatial separation in terms of relieving consumers and farmers from realities of animal work that squeezed profit and removed livestock and their value from the countryside. This distancing made possible cheap food, which also depended on cheap labor in the birthing, breeding, digestive work, and disassembly of the meat-producing animal. The space between the consumer and the meat source allowed for guilt-free, low-cost, high-protein food, which accompanied technological interventions and swelled meat demand and animal supply.[1] The packers and professors at the International helped create and subsequently fill this gulf between animal and human, and they then worked as third parties to rearrange American agricultural life. Consumer and farmer desensitization to the animal experience has been a critical outcome of farm specialization and the International's husbandry reforms, but this emotional detachment has also propelled demand and the dramatic growth in twentieth-century meat consumption.

In the twenty-first century, corn and soybeans along with cattle and calves represent the top farm products annually in the United States.[2] But farmers and feeders largely use corn and soybeans as feed to satisfy consumer protein demand. In 2013, almost half of American corn and over 70 percent of soybeans went to animal feed.[3] Row-crop farmers use commercial fertilizers on largely animal-less farms and then sell corn and soybeans, mostly through intermediaries, to cattle feedlots and high-density poultry and swine facilities. According to a 2015 US Department of Agriculture study, as a percentage of disposable income, Americans paid the least for food relative to other countries. The twenty-first century has seen staggering slaughter numbers for the United States; in 2017, for example, 9 billion chickens, 241.7 million turkeys, 121 million hogs, 32.2 million cattle and calves, and 2.2 million sheep and lambs were killed.[4]

The demand for and supply of chicken grew dramatically following World War II, a phenomenon similar to that seen in the pork industry that had begun relying on high-density feeding and housing structures. The efficiency and technological advancements in chicken production encouraged the dramatic increase in meat production per chicken and rates of gain. In the 1920s, producers could expect a chicken to take as long as sixteen weeks to reach a processing weight of 2.2 pounds. This chicken would have needed 4.7 pounds of feed to produce one pound of meat. By 1993, broilers reached 4.4 pounds in six and a half weeks or forty-five days, requiring only 1.9 pounds of feed to create one pound of meat. Mirroring the knock-on effects initiated by the International in cattle, sheep, and pigs, this change in efficiency dramatically altered the life experience of the animal.[5] Feeders must closely monitor, and even restrict, the grain intake of these modern chickens to prevent them from developing muscle too quickly. This potential for asymmetrical growth has caused some chickens—as well as turkeys—to structurally falter because of muscle growth surpassing the body's capacity to hold the weight, in some cases causing injury and death.

Many of these animals, especially hogs, turkeys, and chickens, are visibly missing from the fields of the American countryside; likewise, meat-producing livestock are also unavailable to farmers and their neighbors. To generate the revenue necessary to service high-cost, monocultural production for row-crop farmers and livestock producers, off-the-farm animal sales have become a necessity. Row-crop farmers have limited choices and control over what to do with their harvested crops. They can either feed the corn and soybeans they grow to livestock or sell them to grain elevators, who then sell much of it to livestock feeders. The only control, it seems, that row-crop farmers have over revenue is through yield, causing them to push for ever-higher yields regardless of whether the market wants more corn and soybeans—a state of affairs encouraged by agronomists, input retailers, and equipment salespeople.

Nonfarm agricultural specialists work alongside reformers and capitalize on industrial husbandry regimes that make farms and rural economies less secure. For many farmers, the need to grow and undercut farmer neighbors has forced them to live on narrow margins. The diseconomies of scale accompanying the "get big or get out" axiom—originally pushed by secretary of agriculture Earl Butz who served under Richard Nixon—have plagued American agriculture for decades. Pushing farmers to increase the size of their operations puts pressure on them to amp up production and to take on unmanageable, risky loans. The very practices agricultural specialists have evangelized—improving yield and increasing gross revenue—have also undermined commodity prices, which led to the debt crisis in the 1980s that drove many farmers out of business and threatened rural livelihoods.

Despite the original goals of the professors at the International to generate more value from fewer animals, the modern animal and the associated crop husbandry regimes have underpinned this capital-intensive, high-yielding version of efficiency that has prioritized volume over value. Relative to corn- and soybean-producers in other countries, American corn and soybean producers bear higher costs, have higher yields, and see higher gross revenues, but they also operate on tighter margins, leaving them economically insecure.[6] Monocultural, specialized farms and farmers, just like their non-farm counterparts, have become inexorably dependent on animal work and on animal bodies leaving the farm, and so they rarely produce the food products they need to survive. This model of extractive agriculture that has developed over the past century has left rural communities, and even their farmers, food insecure. Just as in urban spaces, rural residents live in food deserts too.

★

As objects reconfigured by human desires, the changes in livestock from the late nineteenth century into the twentieth century raises many questions about the concept of animality. Although as biological beings, as scholar Roger Horowitz notes, "animals' bodies resist becoming an expression of our will," they can be altered and exterminated by humans who seek to control biological outcomes.[7] Indeed, reformers dramatically altered the genetics and physiological make up of cattle, sheep, and pigs. But animal influence or even agency did materialize in the consequences of this biopolitical manipulation. Despite the power imbalance between reformers and animals, modern livestock shaped human behavior too, establishing parameters around what was possible and what was needed for them to prosper.

Thinking about the proclivities of modern animals can help inform twenty-first century conversations about the possibility of achieving regenerative agriculture that optimizes food output without destroying local ecologies and exacerbating climate change. The land-grant professors' nineteenth-century criticism of extensive farming and soil robbing still largely applies to commercial farming today, which has not yet become compatible with sustainability. What also emerged from the professors' work was the realization of a serious and ever-increasing demand for food driven by a growing global population, necessitating present-day agricultural scientists, policy makers, and farmers to learn from the role of animal technology in commercial agriculture.

The simultaneous fidelity to technological advancement in agriculture and the growing concern, especially among consumers and activists, about the limits of monocultural crop and animal production highlights an ongoing conflict between a push toward efficiency, first manifested in eugenicist ideas at the turn of the century, and the need for regenerative husbandry in contemporary commercial agriculture. Finding a way to merge regenerative husbandry with commercial agriculture is limited by a history-dependent process that revolved around the construction of modern livestock. Often lost in the conversation on well-being is whether or not livestock have the capacity to be animals, harnessing and benefiting from their own proclivities and diversity, and whether as a result humans too could benefit. The answer is dynamic and ever-changing depending on place, specie, and breed.

The International founders' concerns about food production echo into the twenty-first century. Addressing these questions swirling around food security remains an ongoing challenge that is on display at the livestock expositions scattered across the United States. Local and state governments still fund county, state, regional, and international shows to display what is considered the best animal forms. But underlying these expositions is the requisite obligation for improvement, and so universities, show organizers, and political officials need to discuss what the standards for improvement are and whether those standards are being addressed in the show ring. Equally important to that conversation is what the unintended consequences or knock-on effects of show ring fads are. At the epicenter of this conversation is the North American International Livestock Exposition—the show that carries on the legacy of the Chicago International. In 1971, the Union Stockyards closed after years of decline resulting from the emergence of interstate trucking and the decentralization of meatpacking. The International Livestock Exposition survived a bit longer, until 1975.

The North American International Livestock Exposition filled the void left by the International. Originally called the North American Livestock Exposition, this show takes place in the Kentucky Exposition Center in Louisville,

Kentucky; in its first year, 1974, it featured only beef cattle, but it quickly expanded and soon included sheep, dairy cattle, swine, and quarter horses. As a tribute and successor to the Chicago show, it added the word "international" to its name in 1978. The show adopted many of the traditions of the International in housing the Saddle and Sirloin Club and organizing top collegiate livestock judging competitions. By the show's fortieth anniversary in 2013, it had added several more species, including draft horses, dairy goats, llamas, alpacas, Boer goats, mules, and donkeys. The show also had become the world's largest purebred livestock exposition.[8] With its self-proclaimed name, the organization has unparalleled influence not only in shaping animal bodies but also in creating the standards that college students take back home with them and that influence breeder decisions, just as its predecessor did.

The profit-seeking, business-oriented goals of agribusiness have been undermined in part by the restrictions in biodiversity, adaptability, and immunological and physiological resilience that have arisen from its methods. The moral and economic limitations of this biotic technology can inform real-world agricultural reform, including the work of organizers and judges at state fairs and the North American International Livestock Exposition, in the effort to feed as many as ten billion people by 2050 while also ensuring environmental health and agricultural resilience.

Introduction

1. James "Jimmy" Poole, journalist and expert on the livestock trade, labeled the International the "world's most conspicuous livestock show" ("The Twentieth International"). For more on the International, see also "International Live Stock Exposition." For more on the architectural grandeur of the city, stockyards, and International, see Smith, *The Plan of Chicago*, 57, Miller, *City of the Century*, 199, 318, Pate, *America's Historic Stockyards*, 63–67, 79, "Union Stockyards," *Review of the First International Live Stock Exposition*, 75, and Swift and Company, *The Meat Packing Industry in America*.

2. *Review of the First International Live Stock Exposition*, 10. Many agricultural journals advertised the awe-inspiring built structures and livestock witnessed at the International. See Tormey, "International Just Out of Its Teens," 15. Tormey describes the International as a "Fairyland for the lover of the best in our country's live stock," where one could observe the marvel of "physically perfect animals developed by the master hand" (15). The world-class animals and the newly built grand facilities left no one uncertain about the magnitude of purpose. The International intended, even in its first years, to become the court of last appeal where the best animals were assembled for final judgement. Judging animals, Tormey argues, instilled the "correct ideals of the omega in animal production" (15), which, he adds, benefited the market and consumer, even if indirectly. "The International Swine Show," an account that appeared in the *Swine World*, likewise underscores the spectacle of the event in describing how the "visitor finds a feast for the eyes from the time he enters the arched portals" (3) and also emphasizes the feedback loop established by the exposition. Breeders, breed associations, and people from every agricultural interest converged on the International every year to participate in and witness this great demonstration of progress, the article notes, and then the observers took the lessons learned at the exposition back home or to their professional institutions. The journal *Wool Markets and Sheep* similarly emphasizes this feedback loop created by competition among colleges; institutions that were unsuccessful in the previous year reevaluated their methods and their products in the hope of challenging previous victors ("International Live Stock Exposition").

3. O'Brien, *Through the Chicago Stock Yards*, 26–30. Also see "America to Feed World," which notes that "the food administration designated the International Live Stock Exposition a 'food training camp,'" although the author points out that feeding animals grain to meet these demands did not solve all of the problems: "Never was it so wasteful to feed high-priced grain to ill-bred stock." The International taught producers the combined importance of a feed economy and the use of improved animals to efficiently provide domestic and foreign consumer markets with a reliable, high-quality product. For this reason, the author declares, the food administration recognized "the International as an educational agency."

4. "Union Stock Yards," 6; O'Brien, *Through the Chicago Stock Yards*, 32–34. See also "What Farmers of the Middle States Must Do."

5. "Chicago Selected"; *Review of the First International Live Stock Exposition*; Skinner, "Stock Sales at the International." Skinner helped run the stockyards and served as manager of the International. *Wool and Markets* reprinted a letter he had written that had been mailed to livestock breeders and agriculturalists to advertise the central significance of Chicago in the livestock world. Providing the best prices, the best access through a web of railroads, and "concentration of . . . business," Chicago, he argues in his letter, offered the farmer the greatest economic opportunity. The International, he exclaims, "awakened" the livestock community, prompting the increased production of better, more competitive animals that met the demands of the market. See also James, "Address of Welcome" (1914), box 3, file 44, AHSP; Sanders, "His Influence Upon American Agriculture" (1914), box 3, file 44, AHSP.

6. Several works have explored the relationship between the meat butchered in the stockyards and the consumer; see, for example, Horowitz, *Putting Meat on the American Table*. Others have looked at the politics, the architecture, and the work environment of the stockyards, including Barrett, *Work and Community in the Jungle*, Wade, *Chicago's Pride*, and Slayton, *Back of the Yard*. For an extensive conversation on the impact of railroads and refrigeration on the agricultural and meatpacking industries, see Cronon, *Nature's Metropolis*, and White, *Railroaded*.

7. Many scholars have addressed the stockyard's control over the meat industry. See, for instance, Cronon, *Nature's Metropolis*. Others, such as David Igler in *Industrial Cowboys*, an account of Miller and Lux's large-scale cattle operation that not only butchered animals but raised them, offer a different perspective on organization, treating the ability to supply animals as an essential goal of meatpackers and as a way to mitigate their lack of direct control. The Chicago stockyards, however, could not embrace Miller and Lux's horizontal organization of meat production; the Chicago meatpackers faced a scalability limitation in producing their own supply because they had a much greater demand for meat-producing cattle than the smaller Miller and Lux (in 1913, for example, Miller and Lux's receipts were $5 million, while in the same year the Union Stockyards generated $409 million in revenue).

8. An ad for the International in the *Duroc Bulletin and Livestock Farmer* in 1917 announces the benefit of improved feeding and breeding in meeting the "nation's call" as "a food production camp in the service of the United States."

9. O'Brien, *Through the Chicago Stock Yards*, 32–34.

10. Robichaud, *Animal City*, 5–12.

11. Alexander, *The Mantra of Efficiency*, 2.

12. Woods, *The Herds Shot Round the World*, 3–22.

13. Stern, *Eugenic Nation*, 6.

14. Strom, *Making Catfish Bait out of Government Boys*, xiii–xvi, 1–33.

15. Olmstead and Rhode, *Creating Abundance*, 313–14.

16. White, "Animals and Enterprise"; Jordan, *North American Cattle-Ranching Frontiers*; Skaggs, *Prime Cut*.

17. Specht, *Red Meat Republic*; Anderson, *Capitalist Pigs*; Ritvo, *The Animal Estate*; Anderson, *Creatures of Empire*; Russell, *Like Engend'ring Like*. See also Edmund Russell's *Evolutionary History*. Russell characterizes biological and human events as intertwined; biology, he maintains, should not be separated from political, cultural, and economic developments, and thus he argues for the field of "evolutionary history," or the academic blending of history and biology. This work on "intended evolution" or the explicit manipulation of biological beings through control and breeding is important, but it is also worth noting that these nonhuman actors likewise influence human history; evolutionary history encompasses the give-and-take between biological developments, whether initiated by humans or not, and the experiences and events of human history; it is an ongoing interaction.

18. My study builds on Gabriel Rosenberg's important work on 4-H and the intersectionality of agricultural education and sexuality, *The 4-H Harvest*.

19. Blanchette, *Porkopolis*, 4.

Chapter 1 • *Meatpackers and Professors Take Aim at "Scrubs"*

1. Sanders, *The Story of the International Live Stock Exposition from Its inception in 1900 to the Show of 1941*; Sanders, *A History of Aberdeen-Angus Cattle*; McGee, *On Food and Cooking*, 136; Helmer, *James & Alvin Sanders*. See also the Alvin Howard Sanders Papers at Cornell University, Ithaca, NY, which contain many letters between Sanders and presidents Theodore Roosevelt and William Howard Taft.

2. Sanders, *The Story of the Herefords*, 529.

3. Wentworth, *A Biographical Catalog of the Portrait Gallery of the Saddle and Sirloin Club*, 43–45.

4. Sanders, *Red White and Roan*, 43–45, 246–47, 532–34; Sanders, *Shorthorn Cattle*, 744–45.

5. Olmstead and Rhode, *Creating Abundance*, 264, 286, 323–28; Specht, "The Rise, Fall, and Rebirth of the Texas Longhorn." See also Strom, *Making Catfish Bait out of Government Boys*.

6. Jordan, *North American Cattle-Ranching Frontiers*, 8, 11, 208, 217; Olmstead and Rhode, *Creating Abundance*, 323.

7. Jordan, *North American Cattle-Ranching Frontiers*; Skaggs, *Prime Cut*; Rifkin, *Beyond Beef*.

8. De Loach, *Armour's Handbook of Agriculture*; De Loach, "Beef Cattle." De Loach worked with H. A. Phillips, who was manager of Armour's Sheep Department, to write *Progressive Sheep Raising*. In this publication, they note that the lack of new land available to ranchers forced a change in agriculture. Sheep-raising methods, they argue, needed to shift toward an eastern model. While they acknowledge that sheep still existed on the ranges of the West, they nevertheless maintain that a closed grazing or finishing system required scientific methods that applied to both breeding and feeding.

Accordingly, to raise high-yielding sheep, they insist that purebred breeding and the use of breeds specialized in either wool or meat is necessary for western producers, not just Corn Belt farmers. See also "With Sanders in the Saddle and Sirloin Hall."

9. *Review of the First International Live Stock Exposition*, 164.

10. Worthen argues that the modern animal was "smooth, broad" (*Review of the First International Live Stock Exposition*, 164). Reformers used words like "fat," "broad," and "thick" to differentiate modern animals from range steers that were thin with flat ribs. See also Lambert, *A Trip through the Union Stock Yards and Slaughter Houses*. Lambert illustrates the differences between midwestern cattle and western range steers, arguing that range cattle lacked the overall market appeal of the corn-fed steer and were "generally thin and unfit for cutting into the best grades of meat" (8). Lambert's perspective was echoed in the generalizations made by agricultural reformers who used "range cattle" as a pejorative term to refer to the anti-modern qualities of these animals. See also *Review of the First International Live Stock Exposition*, 163–65.

11. Davenport, "Scientific Farming."

12. Horowitz, *Putting Meat on the American Table*, 1–42.

13. Skinner, "Lifting the Lid."

14. Pate, *Livestock Hotels*, 63–67.

15. Clemen, *The American Livestock and Meat Industry*, 3–91; Hinman and Harris, *The Story of Meat*; Unfer, "Swift and Company"; Hill, "The Development of Chicago as a Center of the Meat Packing Industry"; Knapp, "A Review of Chicago Stock Yards History."

16. Cincinnati meatpackers tried to keep pace with the Union Stockyards by acquiring land, building more pens, and improving production to over five hundred thousand hogs in the 1870s; however, the city's packers could not compete with the technological and strategic advantages that Chicago's networks of railroads and refrigeration capabilities provided the Union Stockyards. See Pate, *America's Historic Stockyards*, Cronon, *Nature's Metropolis*, and White, *Railroaded*.

17. O'Brien, *Through the Chicago Stock Yards*; Grand, *Illustrated History of the Union Stock Yards*; Clemen, *The American Livestock and Meat Industry*, 3–91; Hinman and Harris, *The Story of Meat*; Unfer, "Swift and Company"; Hill, "The Development of Chicago as a Center of the Meat Packing Industry"; Knapp, "A Review of Chicago Stock Yards History."

18. The emergence of the locomotive as primary means of transportation for live cattle eliminated issues of proximity for producers and allowed Chicago to consolidate business operations into one central location. See Wade, *Chicago's Pride*, 51–57.

19. Grand, *Illustrated History of the Union Stock Yards*. The owners equipped the stockyards with the ability to handle fifty thousand cattle, two hundred thousand hogs, thirty thousand sheep, and five thousand horses at one time. Total receipts for 1895 alone reveal how much capacity the facility had: 2,588,558 cattle, 168,740 calves, 7,885,283 hogs, 3,406,739 sheep, and 113,193 horses were brought in. Over the course its first thirty years, the stockyards received nearly fifty million cattle. The stockyards charged producers handling fees for care and maintenance of the animals after arrival. Penning cattle cost twenty-five cents per head, and the feed was an additional cost. The yardage fee helped defray stockyard expenditures, which during this early period ranged from $2,000,000 to $3,500,000 per year.

20. Cronon, *Nature's Metropolis*, 211; O'Brien, *Through the Chicago Stock Yards*, 13–14. For more on the development of the meatpacking industry, see Clemen, *The American Livestock and Meat Industry*, 3–91, Hinman and Harris, *The Story of Meat*, Unfer, "Swift and Company," Hill, "The Development of Chicago as a Center of the Meat Packing Industry," and Knapp, "A Review of Chicago Stock Yards History."

21. Pate, *America's Historic Stockyards*, 75.

22. Wade, *Chicago's Pride*, 51–57; Cronon, *Nature's Metropolis*, 211. See also O'Brien, *Through the Chicago Stock Yards*, Lambert, *A Trip through the Union Stock Yards and Slaughter Houses*, and Grand, *Illustrated History of the Union Stockyards*. In their guidebooks, O'Brien and Lambert take readers through stockyard facilities, illustrating and describing in graphic detail the process of meatpacking. While Grand also details the process of meatpacking, at the same time he offers a biographical perspective, featuring the people and animals well known at the stockyards. In the chapter titled "The Slickest Confidence Game in Chicago," for example, he relates how the packers often used a steer to lead cattle down the chutes to the workers waiting to kill them and used a goat for the sheep. The workers called these animals by the generic name Judas—a reference to the biblical Judas. One "bovine Judas" that workers admired and affectionately remembered, named Phil, worked at the stockyards for five years. They groomed him and blanketed him to protect him from the cold in the winter and flies in the summer. He would roam the yards looking into the cattle pens, and when the drovers notified him, Phil would go to the chute and stand in front. As the cattle filed in behind, he started his walk through the chutes toward the butcher, and then Phil would take a side door out of the chute right before the workers began killing the cattle following him. As he aged, his productivity waned, and one day the side door did not open for him. Phil walked forward under the workers with sledgehammers waiting to begin the butchering process.

23. Lambert, *A Trip through the Union Stock Yards and Slaughter Houses*.

24. "'Bubbly Creek's' Wonders Revealed to Investigators"; "Line Up in War on Bubbly Creek." In 1911, thieves robbed a man and threw the unconscious victim off a bridge into Bubbly Creek. The man floated in the creek for hours and did not drown owing to the presence of semisolid contents in the channel. After he regained consciousness, he struggled to swim out of the substance. Reportedly, it took the man two hours to find his way to shore, "having half climbed and half swum" across the creek. The last few feet of the journey caused the most difficulty, as the creek near the shore was filled with hardened grease. Not only did Bubbly Creek present dangers, but it became an infamous landmark representing waste, dissolution, and damage. Health officials decried the risks of the water by satirically warning that the creek could kill people and could also kill typhoid germs. See also "Bubbly Creek Victim Lives," "Orders Packers to Dig," "Bubbly Creek Dead, but Lives," "Glad Mourners for Bubbly Creek," "Bubbly Creek's Doom Finally Decided Upon," "With a Long Pull and Strong Pull They'll Get Odors from 'Bubbly'," "Ward, Unclean, Kills Babes," and "Yankees Would Swap Rhine for Bubbly Creek."

25. Kujovich, "The Refrigerator Car and the Growth of the American Dressed Beef Industry"; Lawrence, "The Wisconsin Ice Trade." In *The Yankee of the Yards*, authors Louis F. Swift (son of Gustavus) and Arthur Van Vlissingen Jr. reflect on Swift's push for efficiency, his investment in improved productivity, and his drive to further

consolidate business in a central location. Efficient transportation was a concern of many in the industry, one repeatedly voiced in livestock journals, especially the *National Provisioner*, a journal devoted to the American livestock trade. The *National Provisioner* reminded readers that live cattle took up four times as much cargo space as dressed meat. Ideas and articles circulated around efficiency as it related to railways and refrigerated cars, which directly impacted economic competitiveness. Each publication devoted a section to advances in refrigeration. For an extensive breakdown of the risks and costs involved in transporting live animals, the well-being of the animal, and loss of animals in transit, see "Fresh Meats."

26. Pacyga, *Slaughterhouse*, 53–55; Cronon, *Nature's Metropolis*, 234; Miller, *City of the Century*, 207–8. See also Swift and Van Vlissingen, *The Yankee of the Yards*.

27. Miller, *City of the Century*, 208–9; Swift and Company, *The Meat Packing Industry in America*; Cronon, *Nature's Metropolis*, 232–34; "Fresh Meats."

28. *Third Annual Report of the Chicago Junction Railways and Union Stock Yards Company*, 1–13; *Fifth Annual Report of the Chicago Junction Railways and Union Stock Yards Company*; *Sixth Annual Report of the Chicago Junction Railways and Union Stock Yards Company*; "Chicago Tribune"; "Changes on the Range." In 1895, for example, brood cows made up nine hundred thousand head of cattle, or one third of receipts, that were shipped to Chicago (*Fifth Annual Report of the Chicago Junction Railways and Union Stock Yards Company*)

29. *Ninth Annual Report of the Chicago Junction Railways and Union Stock Yards Company*.

30. *Fifth Annual Report of the Chicago Junction Railways and Union Stock Yards Company*; *Ninth Annual Report of the Chicago Junction Railways and Union Stock Yards Company*.

31. The beef shortage troubled meatpackers at the turn of the century. The editorial staff at the *National Provisioner*, the leading journal for the meat trade industry and the official organ of the American Meat Packers' Association, scrambled to address this issue by publishing a series of articles offering solutions to the problem ("For More and Better Meat"). Arguing that the shortage resulted from the closing of the "frontier," decreased livestock numbers, and a vast growth in urban population, A. S. Heath, the author of the series, suggests that improvement of animal quality, not quantity, was the rational way to manage the beef shortage, which would entail the use of "modern" livestock. He goes beyond advocating for judicious breeding and argues that confirmation mattered too, demonstrating that the body of the animal correlated to production and performance. For example, any attempt to fatten a dairy animal for slaughter was fruitless and foolish, because the calories produced went to the production of milk and not meat. See also Nash, "Cattle Paper and the Changing Conditions of the Live Stock Trade," "Chicago Tribune," "Changes on the Range," and "High Prices."

32. "Chicago Tribune"; "Advancing Meat Prices Stir the Press"; "High Prices."

33. Miraldi, *The Pen Is Mightier*; Russell, *The Greatest Trust in the World*.

34. Russell, *The Greatest Trust in the World*.

35. "A Serious Condition." This article tried to gain sympathy among readers. The author gratefully recognizes the efforts of the meatpackers to feed a "nation of meat eaters" and laments anything that might jeopardize the supply—conditions uncontrollable in Chicago—which caused the suffering of many. Meat trade journalists not only

attempted to reverse public opinion with a narrative of service and empathy but also warned that threats regarding the supply of meat or the motives of food distributors, whether they came from a "yellow" journalist or debased politician, were unpatriotic and suicidal. See also Leech and Carroll, *Armour and His Times*, and Armour, *The Packers, the Private Car Lines, and the People.*

36. "High Prices"; Leech and Carroll, *Armour and His Times*; Armour, "The Relation of the Packing House to the Cattle Industry."

37. Armour, *The Packers, the Private Car Lines, and the People*; "Live Stock Show Aid to Industry"; "A Serious Condition"; "The Real Meat Facts"; "Our Beef and Its Maligners"; "Chicago's Food the Best"; "The Meat Situation"; Van Norman, "Live Stock Exchanges and Their Relation to the Producer"; "Chicago Tribune." See also Swift and Company publicity statements and advertisements in the Swift and Company Records, 1879–1954, CHMRC, including "The Missouri Valley Farmer," (1916), box 4, publicity folder, letter to Swift (1916), box 4, publicity folder, letter to Henry Wallace (1916), box 4, publicity folder, letter to the *Christian Herald* (1916), box 4, publicity folder, "There Is No Monopoly" (1918), box 5, folder 1, "High Cattle Prices Accompany High Beef Prices" (1918), box 5, folder 1, "Why Live Stock Prices Go up and Down," box 5, folder 1, "An International Service Built on Tiny Profits Per Pound" (1918), box 5, folder 1, and "You Profit By Our Bigness," (1919), box 5, folder 2, James, "Address of Welcome" (1914), box 3, file 44, AHSP, and Sanders, "His Influence Upon American Agriculture" (1914), box 3, file 44, AHSP.

38. Van Norman, "Live Stock Exchanges and Their Relation to the Producer"; Armour, *The Packers, the Private Car Lines, and the People.*

39. Horowitz, *Putting Meat on the American Table*, 1–42; De Loach, *Armour's Handbook of Agriculture*; Swift and Van Vlissingen, *The Yankee of the Yards*, 69; Cronon, *Nature's Metropolis*, 235.

40. Horowitz, *Putting Meat on the American Table*, 55–74; Pacyga, *Slaughterhouse*, 129; Leech and Carroll, *Armour and His Times*, 321–38; Sotham, "The Potency of Hereford Blood," 343; Armour, *The Packers, the Private Car Lines, and the People*. At the time of the Spanish-American war, the technology was not advanced enough to prevent meat rations packaged for tropical climates from spoiling, giving rise to a "sanitary nightmare for the United States military (Leech and Carroll, *Armour and His Times*, 323). While 345 men died in action in Cuba, Puerto Rico, and the Philippines, over twenty-five hundred died as a result of disease, poor nutrition, and unsanitary conditions—an alarming statistic that provoked backlash toward the meatpacking industry (Leech and Carroll, *Armour and His Times*, 323).

41. "General Review 1901."

42. *Eleventh Annual report of the Chicago Junction Railways and Union Stock Yards Company*; "Stock Breeders' Exposition"; "Unite for a Big Stock Show"; "Breeders' Aid Is Sought."

43. Bowers, *The Country Life Movement in America*, 10. By 1910, farmers represented one third of the US population, and at that time urban populations were booming; in the first two decades of the twentieth century the urban population grew by 80 percent, amounting to twenty-four million additional consumers. Many forces contributed to this shift, including the lure of the city and improvements in transportation and industry and the fact that increased agricultural efficiency lessened the need

for labor on the farm. As the "frontier" closed, the cost of tillable acres rose dramatically, with land prices increasing by an estimated 118.1 percent in the first decade of the twentieth century, leading agricultural experts to shift their focus to increasing efficiency and production and improving crop and animal husbandry practices. Machines and access to off-the-farm inputs increased output and required less labor but also necessitated more capital funds, making farmers more reliant on credit and on the help and services of a network of agricultural experts, many of whom worked at land-grant universities.

44. "The Rural Science Series"; King, *The Soil*; Roberts, *The Fertility of the Land*; Lyon and Fippin, *The Principles of Soil Management*, v–vii; Warren, *Elements of Agriculture*.

45. See Armsby, "The Food Supply of the Future."

46. Bowers, *The Country Life Movement in America*, 9–12.

47. Bowers, *The Country Life Movement in America*, 13–14. For the 1916 income tax return, everyone with net earnings of $3,000 or more was required to file, and when one looks at who filed from each occupation, it is apparent that farmers filed at lower rates. For example, out of six million farmers only a little over fourteen thousand met the threshold that mandated filing. Only one out of every four hundred farmers filed compared to filers in other occupations who had much higher net earnings, including teachers, ministers, salesmen, doctors, and engineers.

48. *Report of the Commission on Country Life*, 37. Even though Roosevelt felt that the farmer was better off in 1908 than ever before, he was still concerned that rural life lagged behind urban society. In outlining the goals of the County Life Commission to Bailey, he laments that the farmer's "increase in well-being has not kept pace with that of the country as a whole" (41). The president wanted Bailey, as chairman of the commission, to address this disparity in development and to improve methods of farming to end the "suffering and needless loss of efficiency on the farm" (41). For Roosevelt, better farming was key to achieving broader societal goals; increased productivity and efficiency in animal and crop husbandry would, he believed, provide a better lifestyle for farmers, and improving the income of farmers through better marketing and business practices combined with greater yields would directly address deficiencies in rural life. See Curtiss, "The Rural Education Problem" (1911), box 1, folder 9, CFCP, Bailey, *The Country-Life Movement in the United States*. See also Christie, "The New Agriculture" (1916), MSF 89, folder 2, GICP, Christie, "Opening Address" (1913), MSF 89, folder 1, GICP, Gross, "Better Country Life" (1913), MSF 89, folder 1, GICP, and Waters, "The Building of National Agricultural and Country Life" (1913), MSF 89, folder 1, GICP.

49. "Suggestions for Keeping Boys on Farms." These commentators offer many solutions, including a nicely-bred, well-organized team of draft horses, like Percherons. A good team of horses decreased the amount of physical labor needed and allowed farm children to concentrate on the skilled labor of running the machine, perfecting animal and crop husbandry practices, and using progressive management skills to operate the farm. See also "How to Make Farm Life More Attractive," (1908), box 21, folder 1, LHBP, Warren, *Elements of Agriculture*, Cook, "City and Country," and *Rural Communities and Centers of Population*, box 24, folder 3, LHBP.

50. Adams, *The Modern Farmer in His Business Relations*; Shaw, *Animal Breeding*; Hopkins, *Soil Fertility and Permanent Agriculture*; Hopkins, "Plant Food in Relation to

Soil Fertility"; Mumford, *Beef Production*; Lyon and Fippin, *The Principles of Soil Management*; Henry and Morrison, *Feeds and Feeding*; Plumb, *Beginnings in Animal Husbandry*; Plumb, *A Study of Farm Animals*; "Barnyard Manure"; Winkleman, "Problems of American Agriculture"; "To Maintain Soil Fertility"; "Two Methods of Farming"; "System of Robbing the Soil"; "Decline in Rural Population"; "Urban vs. Rural Population"; "Hopkins Addresses Rural Teachers"; "A Nation of Soil Robbers"; "The Biggest Problem on the American Farm"; "How to Keep Boys on the Farm"; "Maintenance of Fertility"; "Exhausting the Soil"; Hopkins, "The Story of the Soil"; "Decline in Soil Fertility"; Golloway, "Soil Fertility"; Sanders, "Cattle as a Major Factor in Successful Farming," box 4, file 32, AHSP; Sanders, "Hobbies," box 4, file 32, AHSP.

51. Adams, *The Modern Farmer in His Business Relations*, 48–69. See also Warren, *Elements of Agriculture*, "Science and the Farmer" (1893), vol. 2, CSPP-OSU, and Plumb, "Free Literature for Progressive Farmers," vol. 3, CSPP-OSU. Plumb, who was a professor at Ohio State University, linked good or productive farming to intelligent inquiry and scientific crop and animal husbandry regimes, what he saw as the core tenets of modern farming. Criticisms directed at professors for being overly theoretical or bookish were, he admitted, valid to an extent, but even if the prejudice against book farming was partially justified, he believed the twentieth century demanded reconciliation between the farmer and the emerging science produced by land-grant schools. The principles of science tested at experiment stations, the results of which were distributed by these universities, focused on practical applications and methods. Like his colleagues, Plumb framed soil exhaustion as the central concern for agricultural scientists, and in these reviews of agricultural reform, he argues that "intelligent farming" was the only cure for the ills of nineteenth-century husbandry practices.

52. Adams, *The Modern Farmer in His Business Relations*, 69.

53. Lyon and Fippin, *The Principles of Soil Management*, xxx–xxxi. Cook, "City and Country."

54. "Two Methods of Farming"; Adams, *The Modern Farmer in His Business Relations*; Shaw, *Animal Breeding*; Hopkins, *Soil Fertility and Permanent Agriculture*; Hopkins, "Plant Food in Relation to Soil Fertility."

55. Sanders, "Cattle as a Major Factor in Successful Farming"; Warren, *Elements of Agriculture*; Sanders, "Hobbies," box 4, file 32, AHSP.

56. Roberts, *The Fertility of the Land*.

57. Schlebecker, *Cattle Raising*, 11; Jordan, *North American Cattle-Ranching Frontiers*, 7–13, 210–17. See also Roberts, *The Fertility of the Land*, Hopkins, *Soil Fertility and Permanent Agriculture*, xvii–xxiii, 198–235, Warren, *Elements of Agriculture*, and Short, "Ancient and Modern Agriculture," box 3, file 43, AHSP.

58. Hopkins, *Soil Fertility and Permanent Agriculture*, xvii–xxiii, 198–235. Hopkins insisted that finding the proper balance between commercial fertilizer and animal manure was essential for permanency. Hopkins was not as enamored with the idea of using livestock as machines for human food production as his land-grant colleagues; he worried about the caloric loss in the conversion of crops to meat and argued that perhaps it would be easier to meet the food needs of humans by avoiding dependence on a meat-based diet. Despite his uncertainty about livestock and meat production, he adamantly believed that permanent agriculture depended on livestock manure. Hopkins also cited recent research conducted at experiment stations in Tennessee and

North Dakota that showed problems with weeds, insects, and fungus in one-crop systems. Disease, competing plants, and pests plagued monocrop farms and diminished yields, farm productivity, and revenue.

59. Roberts, *The Fertility of the Land*; Hopkins, *Soil Fertility and Permanent Agriculture*, xvii–xxiii, 198–235; Warren, *Elements of Agriculture*; Short, "Ancient and Modern Agriculture," box 3, file 43, AHSP; Christie, "Supplying the Farm Labor Need" (1918), MSF 89, folder 2, GICP.

60. Armsby, "The Food Supply of the Future."

61. Poole, "The Twentieth International."

62. Sanders, *At the Sign of the Stock Yard Inn*; Sanders, *Story of the International*, 25–30; "Dedication of the New Building"; *Ninth Annual Report of the Chicago Junction Railways and Union Stock Yards Company*; "Live Stock Exposition"; "Stock Show to Be Great"; "Stock Breeders' Exposition"; "Unite for a Big Stock Show"; "For Chicago Live Stock Show"; "Breeders' Aid Is Sought"; Union Stock Yard and Transit Company minutes book, 1865–1904, CHMRC.

63. Sanders, *At the Sign of the Stock Yard Inn*, 8–22; Sanders, *The Story of the International Live Stock Exposition*, 25–30; "International Live-Stock Exposition"; "Live Stock Exposition"; "Unite for a Big Stock Show"; "Breeders' Aid Is Sought"; "Dedication of the New Building"; Plumb, "International Live Stock Exposition" (1900), vol. 3, CSPP-OSU.

64. Sanders, *At the Sign of the Stock Yard Inn*, 8–22; Sanders, *The Story of the International Live Stock Exposition*, 25–30; "Unite for a Big Stock Show"; "Breeders' Aid Is Sought."

65. Sanders, *At the Sign of the Stock Yard Inn*, 8–22; Sanders, *The Story of the International Live Stock Exposition*, 25–30; "Unite for a Big Stock Show"; "Breeders' Aid Is Sought."

66. Curtiss's organizational and associational work addressed the value of purebred livestock, nutrition and feeding practices, and general agricultural improvement. He first started at Iowa State Agricultural College as a student and received his degree in 1887. He returned to Iowa State College after a few years as a farm manager. Curtiss worked with James Wilson to build a robust agriculture department. After Wilson went to Washington, DC, to serve as secretary of agriculture, Curtiss became the director of the program. In 1900, Curtiss was promoted to dean. On Rookwood Farm, he raised and exhibited his own livestock, and producers recognized him for his frequent judging engagements at the International as well as other high-stakes expositions, including the Panama-Pacific Exposition, and nearly all leading state and provincial fairs in the United States and Canada. He was also sought after as a gubernatorial candidate in the Iowa Republican Party and was rumored to have been seriously considered for secretary of agriculture under President Calvin Coolidge. See Campbell, "He Put the 'A' in Agriculture"(1952), box 1, folder 8, CFCP, Beckman, "Dean C. F. Curtiss," (1922), box 1, folder 8, CFCP, Randolph, "C. F. Curtiss Dies," (1947), box 1, folder 1, CFCP, and Pammel to Coolidge (1924), box 1, folder 8, CFCP.

67. Sanders, *At the Sign of the Stock Yard Inn*, 8–22; Sanders, *The Story of the International Live Stock Exposition*, 25–30.

68. Poole, "The Twentieth International." Poole notes a rapid decline in the use of inferior bulls and "scrub females" in both show herds and flocks as well as commercial

operations over the course of two decades and provides an interesting take on how the International influenced this change. He argues that the International educated and demonstrated to college students and active producers the value of improved stock, which shifted preferences and resulted in "inferior" or scrub animals losing their "footing under the influence of accumulating hostile public opinion" (14). He observes that by changing public opinion, the International institutionalized "progress of reform" (14) in American agriculture, adding that the International was created to "carry on the perennial battle for the extermination of the scrub" (13). The International, Poole notes, marked a turning point in the production of livestock in North America, and the exposition provided order in the agricultural community out of the "gigantic hodgepodge" and "veritable turmoil" (13) of the previous century. See also Plumb, "The Use of Words by Breeders" (1901), vol. 4, CSPP-OSU.

69. Sanders, *At the Sign of the Stock Yard Inn*, 8–22; Sanders, *The Story of the International Live Stock Exposition*, 25–30.

70. Sanders, *At the Sign of the Stock Yard Inn*, 8–22; Sanders, *The Story of the International Live Stock Exposition*, 25–30, 37; "Live Stock Show Opens Gates Today: International Exposition, Attended by Stock Fanciers from All over the World, Attracting More Attention Than Ever This Year" (1909), box 2, file 38, AHSP; "Alvin Sanders" (1909), box 2, file 38, AHSP. See also Plumb, "International Live Stock Exposition," in which he proclaims the first International an overwhelming success. He estimated that fifty thousand spectators passed through the gates each day, and the hotels hardly had a vacant room, requiring excess travelers to sleep on cots in dining areas and halls. The high number of participants, tourists, and spectators spurred growth in the local economy, and so immediately following the conclusion of the first show, the organizers began planning the 1901 International.

71. Sanders, *At the Sign of the Stock Yard Inn*; "Dedication of the New Building"; "Largest Exposition Building in World Nearing Completion in Chicago"; "Praises Show of Livestock."

72. "New International Live Stock Exposition Building"; "The Coming International"; O'Brien, *Through the Chicago Stock Yards*, 30–32. In 1905, the board of directors postponed the International two weeks to give contractors the needed time to finish construction. The delay resulted from late shipments of structural iron. Feeders planned their feeding and marketing on the original date and complained that the delay would cause them to miss out on the high prices of early December stimulated by eastern buyers preparing for the coming holiday season. W. E. Skinner, the general manager, attempted to placate producers and potential spectators, stating that, despite delays, they could expect to see "the finest bovine aristocracy" ("The Coming International").

73. Davenport, "Scientific Farming."

74. "With Sanders in the Saddle and Sirloin Hall"; Poole, "The Twentieth International."

Chapter 2 • Breeding Up Livestock

1. See *Opportunities of To-Day* 's "International Live Stock Exposition." According to *Opportunities of To-Day*, the dedication of the of the Purebred Livestock Record Building marked the beginning of the improvement of livestock resources in the United

States, an improvement that was sustained by the yearly exposition, which as "the exponent of a great movement for improvement of the domestic animals of the United States" had had a substantial impact in just a few short years on the choices of producers through the awards, competitions, and education it offered.

2. "International Live Stock Exposition." For Wilson, one of the major benefits of the exposition was the education it offered to college students. The standards learned by students, especially in the popular collegiate judging competition, were widely disseminated and influenced the types of livestock producers raised for other shows. When they graduated and judged other shows, these college alumni prioritized the type of animal they learned about at the International. The value of college students "scoring" animals at the International related to the general improvement of agriculture, which, Wilson suggested, provided an unprecedented opportunity for Americans, "as a people," to go to the front of other nations in "the production of meats."

3. "Union Stock Yards."

4. Wentworth, *A Biographical Catalog of the Portrait Gallery of the Saddle and Sirloin Club*, 8–9. Wentworth worked for Armour and Company as an executive, and he also worked at the International. As the ringmaster for the International, he served as the face and voice for the its proceedings. In 1903, Wentworth declared that the International had quickly emerged as the hub, or "pivot," for agricultural improvement. He saw the Saddle and Sirloin Club as an important institution in establishing and commemorating the forgotten British producers who had initiated this movement. British-born Richard Gibson, who bred improved cattle in North America, was the one who came up with the name; he borrowed it from writer H. H. Dixon, who writing under the pen name "the Druid" had published a volume dealing with British breeders titled *Saddle and Sirloin* ("saddle" referred to top cuts of meat on the sheep and "sirloin" to those of the cow).

5. Sanders, *At the Sign of the Stock Yard Inn*, 3–50, 155–60; Wentworth, *A Biographical Catalog of the Portrait Gallery of the Saddle and Sirloin Club*, 7–10; *Saddle & Sirloin Portrait Collection Guidebook*; "Saddle and Sirloin Club (unedited)," box 1, folder 1, ILER; "Saddle and Sirloin Club (edited)," box 1, folder 1, ILER; "Special Release from International Livestock Exposition Press Bureau," box 1, folder 1, ILER; "Statement of Incorporation of the Saddle and Sirloin Club of Chicago," box 1, folder 1, ILER; Henkle, "Saddle and Sirloin Club," box 1, folder 1, ILER; "With Sanders in the Saddle and Sirloin Hall"; "Cattle as a Major Factor in Successful Agriculture," box 4, file 32, AHSP; Plumb and Hum, "History of the Saddle and Sirloin Club," vol. 4, CSPP-OSU; "Prize Stock Given Finishing Touch;" *A Review of the International Live Stock Exposition*, 1913, 292, 294.

6. *A Review of the International Live Stock Exposition*, 1913, 292, 294.

7. "Robert Bakewell"; "Charles Robert Darwin"; "Gregor Johann Mendel."

8. Sanders' devotion to scientific breeding is not only evident in his work as a writer and editor for the *Breeders' Gazette* but also in several publications about the benefits of purebreds: *A History of Aberdeen-Angus Cattle*, *Red White and* Roan, and *The Story of Herefords*. In 1900, he published *Short-Horn Cattle*: in which he examines the history of genetic selection and famous breeders, such as Robert Bakewell, who provided producers guidance in phenotypical improvement through purebred influences.

9. "The Gospel of Improvement"; Sotham, "The Potency of Hereford Blood."

10. Campbell, "He Put the 'A' in Agriculture" (1952), box 1, folder 8, CFCP; Beckman, "Dean C. F. Curtiss" (1922), box 1, folder 8, CFCP; Randolph, "C. F. Curtiss Dies" (1947), box 1, folder 1, CFCP; Pammel to Coolidge (1924), box 1, folder 8, CFCP. The Charles F. Curtiss Papers in the Iowa State University Library Special Collections Department contain an interview with the professor, undated, that details his broader optimism about the agricultural industry if the Corn Belt model of animal husbandry were to become more dominant. Curtiss emphatically reminds the interviewer that livestock on every farm was necessary for farms to maintain or improve their value because of the fertility benefits garnered from manure. Without a continued commitment to fertility, which he believed required animals, the nation as a whole would fail to meet the food demands of the growing consumer class.

11. Spoor, "Tells of Great Year." Spoor explains why he rejects the scrub, noting that its deficiencies resulted not only from genetics but also feeding regime. Spoor notes that animals needed to be fed according to scientific standards, as genetically improved livestock raised properly captured a premium at the market. In Spoor's view, purebred cattle, especially Herefords, Angus, Shorthorns, and Galloways, possessed the genetic and practical traits that made it possible to consistently produce prime beef whether the cattle were raised on the western range or a smaller farm in the Midwest. See also "International Exposition," which declares that the success of the International doomed the scrub animal, as the demand for "fine" animals grew as a result. Better selection and demand drove the sale price for improved stock, which greatly outpaced the value of the average animal. In 1901, western and range cattle sold in Chicago faced a weakening market. In contrast, the market remained strong for improved animals of the Midwest both for show and commercial purposes. See "Cattle."

12. American reformers pushed the importation of purebred stock from Britain; in return the United States sold large numbers of commercial animals to Britain. The trade from 1870 to 1899 generated a sale of 389,490 cattle, 143,286 sheep, 45,778 horses, and 6,775 mules to Britain at total value of nearly $38 million (see Plumb, "Crossing the Atlantic: University Men in Charge of Horses, Cattle and Sheep Aboard Tritonia—Ship's Equipment and Incidents of the Voyage" [1900], vol. 2, CSPP-OSU, and *Wool Markets and Sheep*'s two-part "British Sheep Farming"). For Britain as an island nation, vulnerable to blockade, both domestic production of food and the ability to import it was essential. Even more, meat in particular offered the British a sense of general superiority. Britain's elite animals encouraged its citizens to see Britain as a great nation and supplier of great genetics to the world, although there were conflicts between the idea of service and the realities of elite livestock production, as breeders raised animals with aesthetic appeal in mind, not commercial production. For more on the intersections of patriotism and agriculture in Britain, see Ritvo, *The Animal Estate*, 46–52.

13. De Loach, *Armour's Handbook of Agriculture* ("Beef Cattle"). De Loach worked for Armour, and he believed that importing British livestock and increasing its presence on the average farm would address many of the packers' supply concerns.

14. Wentworth, *A Biographical Catalog of the Portrait Gallery of the Saddle and Sirloin Club*, 154–56.

15. McFarlane, *The American Aberdeen-Angus Herd-Book*. See also Plumb, "Purebred Live Stock Associations and Their Methods," pt. 6 (1907), vol. 4, CSPP-OSU. The exposition provided the space for breed societies to sell the best their breed had to

offer. These sales, Plumb comments, were not necessarily open to all breeders; instead, agents of the associations often selected animals to be entered into the sale to ensure top quality livestock, which avoided the "scalawag type." For more on the role of breed associations in livestock improvement, see "The Gospel of Improvement," *Chicago Livestock World*.

16. Plumb, "International Live Stock Exposition" (1900), vol. 3, CSPP-OSU. The *Drover's Journal* awarded the champion steer with a decorative cup. Following the show, Advance weighed 1,430 pounds at the auction and sold for $1.50 per pound, which was the highest price ever paid for a steer to date, according to Plumb. Breeders showed Aberdeen-Angus cattle in large numbers at the first International, which helped promote purebred animal husbandry. The International mimicked the Columbian Exposition, creating a show that featured the changing components of American agriculture and demonstrating to the public and producers what modern animal husbandry looked like. The International can thus be thought of as a world's fair of the agricultural community; it both presented the best animals of the period and served an educative function by disseminating ideas and persuading producers to adopt new practices. For a broader discussion on world's fairs, see Rydell, *World of Fairs*, and Rydell, *All the World's a Fair*.

17. *Review of the First International Live Stock Exposition*, 23–31.

18. US Department of Agriculture, "Statistics of Agriculture"; Coulter, "Agricultural Development in the United States, 1900–1920."

19. See the advertisements in the *Review of the International Livestock Exposition* albums and the *Live Stock Journal Almanac* publications from London. The White Star Line, for example, advertised convenient shipment dates and competitive cargo fares for shipping livestock between Liverpool and New York every Friday on the "steamers." These advertisements assured potential clients that these ships, which had electric lights and a water supply, were designed specifically for the safe and comfortable travel of blooded stock. Companies also provided care for the livestock on these ships, hiring a "surgeon" or animal expert. See *Live Stock Journal Almanac*, 349.

20. Curtiss, "An American on British Stock."

21. Curtiss, "An American on British Stock." See also Plumb, "An American Agriculturalist Abroad" (1897), vol. 2, CSPP-OSU, "Across the Atlantic with Live Stock" (1897), vol. 2, CSPP-OSU, and "An Agriculturalist Abroad" (1900), vol. 2, CSPP-OSU, in which Plumb recounts details of his trips to England and Scotland to observe agriculture and study their stock shows. He notes that British consumers, in contrast to American ones, preferred mutton, and so meatpackers specialized in lamb slaughter and also remarks on how sheep grazing there was different from that in the United States. British sheep were bred to utilize the foliage, and the animals balanced crop regimes because they needed very little grain during the summer months to perform well and they also returned fertilizer to the soil. He also wrote a piece on how British breeders raised and selected animals specifically for a purpose, which resulted in a uniformity among British purebreds, breeds that he thought provided great value and opportunity for the American farmer ("Sheep Husbandry in England and Scotland" [1907], vol. 2, CSPP-OSU). Uniformity in type, Plumb notes, prevailed to such a degree in Britain that even crossbred animals bore similar traits and standards as the purebreds, such as broad backs, deep bodies, short legs, and good wool.

22. "The Court Decision Governing Importations of Breeding Stock."

23. For the decision, see "United States v. One Hundred and Ninety-Six Mares." See also Abott, *A Digest of Reports of the United States Courts*, 239, and "The Court Decision Governing Importations of Breeding Stock." The editors and writers at *Breeder's Gazette*, especially James and Alvin Sanders, however, objected to the court's protecting only individual farmers and not middlemen as well from the customs duty. They argued that the distinction between the producers importing animals for breeding purposes and those who found animals of good character and value to import and sell to farmers was a damaging and arbitrary one that imperiled American agriculture. They maintained that the court needed to consider the intent of the law, which was to provide for "the free entry of animals especially imported for breeding purposes . . . to encourage improvement in live stock."

24. Cooper, "The Export Trade in Pedigree Stock"; "British Sheep Farming."

25. "The Babraham Southdown Flock," 1901, 332.

26. "The Babraham Southdown Flock," 1910, 306.

27. *A Review of the International Live Stock Exposition*, 1913, 245.

28. Tunis were often referred to as broad-tailed sheep owing to their fat tails, which store nutrients like a camel's hump. William Eaton first shipped Tunis to the United States in 1799. Eaton served as consul for the United States in Tunis, and he received permission to ship ten Tunis on the man of war *Sophia*. The sheep that survived the trip found a home at the farm of Judge Richard Peters near Philadelphia. Peters raised Tunis for twenty years, and their mutton drove demand for the breed in the Philadelphia area. The high rate of slaughter that this demand entailed hindered the development of the breed. Many with government affiliations used their positions to leverage the importation of Tunis for their own flocks, including Commodore Barron of the United States Navy and Thomas Jefferson. Flocks spread throughout the North and the South, but they became obsolete in the North in subsequent decades because of the craze for the high-quality fine wool produced by Merino sheep. When American breeders switched back to mutton breeds, they replaced their flocks with British breeds. In the South, nearly all the Tunis flocks were wiped out by the Civil War. A surviving flock from South Carolina helped reestablish the breed in the 1890s. This flock participated at the 1893 World's Fair in Chicago. Charles Roundtree, the breed's chief promoter, purchased Tunis from this southern flock and began to rekindle the breed in the North with J. A. Gulliams. Both Roundtree and Gulliams lived in Indiana, and they started the American Tunis Sheep Breeders' Association in Fincastle, Indiana, which later moved to nearby Crawfordsville. The association worked to advance the breed and advertise the benefits of the African broad-tailed sheep in the United States, and by 1914 the association registered 2,530 Tunis sheep. Nevertheless, Tunis did not have a place at the International; the preference for British sheep remained strong. See Roundtree, "Why I Breed Tunis Sheep," *The Annual Register of a View of the History, Politics, and Literature, For the Year 1810*, 624–30, "The Foreign Breed of Sheep," 124–125, Shaw and Heller, "Domestic Breeds of Sheep," 30–33, "A Sketch of President Guilliams and His Tunis Sheep," Connor, "A Brief History of the Sheep Industry in the United States," and Plumb, *Types and Breeds of Farm Animals* 425–28.

29. In the first issue of its house organ, the association outlined its founding principles by introducing Darwin and Mendel. Darwin's *Origin of Species* was suggested

reading for farmers in many of the progressive farming texts of the period. See Adams, *The Modern Farmer in His Business Relations*, and Miles, *Stock-Breeding*.

30. Williams, "Eliminating Hazards in the Range-Cattle Business," Poole, "Beef Production on New Basis."

31. Stern, *Eugenic Nation*, 16–17.

32. Romo, *Ringside Seat to a Revolution*, 240–43; Richardson, *How the South Won the Civil War*, xxv.

33. Levario, *Militarizing the Border*, 1–16.

34. Stern, *Eugenic Nation*, 21.

35. Stern, *Eugenic Nation*, 231.

36. Plumb, *A Study of Farm Animals*, 220.

37. Richardson, *How the South Won the Civil War*, xxii–xxiii.

38. Stern, *Eugenic Nation*, 109, 157; Specht, *Red Meat Republic*, 6, 7–8, 21–66.

39. Strom, *Making Catfish Bait Out of Government Boys*, xiii–xvi, 1–33; Mumford, *Beef Production*, 138–42; Olmstead and Rhode, *Creating Abundance*, 322–23, 399–400.

40. Strom, *Making Catfish Bait Out of Government Boys*, 2, 10, 17–19.

41. Strom, *Making Catfish Bait Out of Government Boys*, 30–31.

42. Strom, *Making Catfish Bait Out of Government Boys*, 41.

43. Strom, *Making Catfish Bait Out of Government Boys*, 3.

44. In *The War Against the Weak*, Edwin Black argues that American political structures adopted eugenics more quickly than European counterparts. The US Department of Agriculture endorsed the "master race" ideology and hereditary selectiveness, and associations devoted to eugenics emerged to support and disseminate these ideas in the hopes of mitigating many of the anxieties that plagued American society. See also Glenna, Gollnick, and Jones, "Eugenic Opportunity Structures," and Kimmelman, "The American Breeders' Association."

45. Jordan, "Report of the Committee on Eugenics"; Bell, "A Few Thoughts Concerning Eugenics."

46. Jordan, "Report of the Committee on Eugenics"; Bell, "A Few Thoughts Concerning Eugenics." In addition to being a famous inventor, Bell was also a sheep breeder, and he investigated the genetic transmission of certain traits, like the number of nipples on sheep.

47. Van Wagenen, "Preliminary Report of the Committee of the Eugenics Section of the American Breeders' Association to Study and to Report on the Best Practical Means for Cutting Off the Defective Germ-Plasm in the Human Population." In this presentation to International Eugenics Congress, Bleecker Van Wagenen, chairman of the eugenics section of the American Breeders Association, discussed how defects are inherited and speculated on the potential to eliminate "anti-social" and "defective traits" in the human population. Those with such traits, he declared, constituted a "drag on society" and handicapped "industrial and social" progress. He advocated for sterilization, both compulsory and voluntary, to eliminate unwanted traits as well as the segregation of targeted individuals, at least during their reproductive years. He also recommended restrictive marriage laws, education regarding the inheritability of negative and positive traits, improvement of environmental conditions, and euthanasia. He mentioned polygamy as a way to manage reproduction and artificial interference as a means of preventing conception. Finally, he argued that creating systems of reproduc-

tion to "remove defective traits" could be useful. He lauded Indiana, especially the Jeffersonville Reformatory, and California for passing laws supportive of sterilization, but he cautioned that sterilization as a practice for improvement among the human population required patience and foresight. Most of the sterilized, he argued, would still be more or less "dependent" on society, and those "afflicted" with social misbehavior, especially sexual immorality, would continue to behave as such. However, Van Wagenen believed that society would begin to see benefits generationally as states prevented the reproduction of "unworthy progeny." When Van Wagenen made his case to the congress, he likened the genetic culling of human genes to the selection of desirable livestock.

48. "The Woman Movement and Eugenics." This author was primarily concerned that women of the best financial, physical, and moral fitness were occupying their time with education, social events, politics, and work instead of contributing to racial uplift through childbearing—a part of these biologically elite women's "racial duties" (226). This author urges readers to learn from plant and animal breeders and apply the lessons of modern animal husbandry. The "best" blooded people needed to reproduce at high rates, "medium" blooded should reproduce but more cautiously, and people of "low" blood should not reproduce (227). To accomplish this, eugenicists hoped to collect data on humans to determine their eugenic value, just as they did with pedigreed animals. In this way, the author hoped, the American Breeders Association could encourage the development of "a race of high average sanity, health, and general efficiency" (228). See also "Race Genetics Problems," in which the author correlates racial purity within "aryo-germans" (231) with the advancement of civilization on the American continent, arguing that the political, commercial, and industrial successes of the "long-light, blond, long-skulled race" (231) did not result from race mixing. Despite the seemingly satirical tone, the editorial suggests that white people's own inventiveness in conquering many regions of the world facilitated the mingling of the races, which threatened their own racial purity. To address this problem, eugenicists clamored for further investigation into the deleterious impact of the "mingling" of races and the benefits afforded society by "the superiority of pure racial stocks" (232). The push for research into beneficial or dominant positive traits and the undesirable traits of human races was similar to the call for investigation of purebred livestock.

49. Plumb, *A Study of Farm Animals*, 21–47. For more information, see Miles, *Stock-Breeding*, and Marshall, *Breeding Farm Animals*. Davenport devoted much of his work and publications to heredity, genetics, and eugenics; see *Heredity in Relation to Eugenics*, in which he outlines the general principles of heredity and the mitigation of human problems through genetics. Immigrants, he maintains, should be selected based on blood: "In other words, immigrants are desirable who are of 'good blood'; undesirable who are of 'bad blood'" (222). He proposes ways to execute this policy, including through the collection of genetic information and the use of American agents to scout for immigrants in Europe. He also argues for going beyond simply aiming to eliminate unwanted traits, urging eugenicists to learn from crop and animal farmers about the positive breeding or pairing of mates to increase the likelihood of breeding better species: "Proper matings are the greatest means of permanently improving the human race—of saving it from imbecility, poverty, disease and immorality" (260).

50. Plumb, *A Study of Farm Animals*.

51. "Plumb, Charles, S.," MSF 312, folder 2, CSPP-PU; "Charles Sumner Plumb," MSF 312, folder 6, CSPP-PU. See also Plumb, "The Care of Domestic Animals," MSF 312, folder 4, CSPP-PU.

52. Plumb, "The Function of the Constructive Breeder of Registered Live Stock." Plumb urged farmers to buy or raise the "right type" of females and also to select males to correct deficiencies. An "improved" male had a larger net impact on the farm since he genetically represented half of the herd or flock, and the selection and use of a "well-bred male" even if that individual animal came at a greater cost to the farmer. The investment made in a single great male injected value into the entire herd or flock, which resulted in a net gain for the producer. Plumb likened the buying of cheap, "inferior" animals to speculation, which burdened farmers with risk and threatened their farm's general productivity and potential profitability. However, one type of animal did not suit all production systems. Thus, "improved" animals did not all look alike or come from the same purebred heritage, and once cattle producers, for example, decided what commodity they wanted to sell, then they needed to identify and select animals with body types and genetics suited for the specific production of that commodity. In this way, Plumb urged producers to embrace specialized production not just on their farms but also in the physiological and genetic makeup of their livestock. The specialization of animal type correlated with increased productivity, efficiency, and quality of product, and for Plumb, these were the central elements of constructive breeding.

53. Plumb, "The Function of the Constructive Breeder of Registered Live Stock."

54. Ritvo, *Animal Estate*, 68; Russell, *Like Engend'ring Like*.

55. Miles, *Stock-Breeding*, 2–10; Ritvo, *Animal Estate*, 68; Russell, *Like Engend'ring Like*. The desire for noble or "pure" breeding encouraged the institutionalization of British breed societies and the recording of animal ancestry.

56. Farmers used inbreeding to improve the consistency and status of offspring; however, not only did they face ethical objections but they observed negative consequences, including lethargy, weakness, and low rates of fertility and vigor, which impacted the overall health of the animals on the farm. For further information, see Russell, *Like Engend'ring Like*, which notes that inbreeding was also deceptive in that it could create a perception of quality in pedigreed stock where it did not exist. Russell also describes approaches that avoided excessive inbreeding, including hybrid stock breeding (selecting two different purebred animals and mating them in an attempt to combine the positive attributes of two purebreds and avoid the weaknesses involved with close breeding, an approach that led to wholly new breeds of livestock in the United States) and the importation of purebred stock into regions or areas that needed new genetics, through which over generations a local breed could be turned into a "foreign" breed.

57. Lloyd-Jones, "What Is a Breed?"

58. Ritvo, *The Animal Estate*, 60–68. For examples in Hereford breeding, see Sotham, "The Potency of Hereford Blood," Marshall, "A Study in Scotch Pedigree," and "The Gospel of Improvement."

59. For more discussion on the negative consequences of inbreeding, see Kraemer, "Effects of Inbreeding." In this article, Kraemer, a German academic, argues that breeders must reject the notion that inbreeding is harmless, citing many studies,

including the work of Darwin, to navigate the complexities of inbreeding and the potential health problems it gave rise to. Kraemer explores the question of whether inbreeding as such was bad or if rather only inbreeding from "inferior" stock was and concludes that the problem was inbreeding itself: "Continued inbreeding always must result in weakened constitution, through its own influence" (234). Although he is unwavering in his rejection of inbreeding or close breeding, he maintains that all improved breeds developed from intensive linebreeding and that students of improved agriculture need to understand the developmental influence of close breeding in the creation and standardization of prominent breeds of livestock.

60. Plumb, *Types and Breeds of Farm Animals*, revised ed., 657–59.

61. Marshall, *Breeding Farm Animals*; Miles, *Stock-Breeding*, 138–40.

62. Marshall, *Breeding Farm Animals*, 185. For a broader conversation on close breeding, see Marshall's chapter "Inbreeding and Line Breeding" in *Breeding Farm Animals* and Miles's chapter "In-and-In Breeding" in *Stock-Breeding*.

63. Marshall, *Breeding Farm Animals*, 73–78, 188–208. Not all unlikely, atavistic, or even mutated traits possessed a negative value. Many purebred animals created new breeds, such as the Polled Durham, which resulted from a mutation in Shorthorns that eventually eliminated the horn. By mating those unusual traits with animals with similar inclinations, breeders changed animal types, aesthetics, or even purpose over time.

64. Plumb, *A Study of Farm Animals*, 21–47.

65. Marshall, *Breeding Farm Animals*.

66. Marshall, *Breeding Farm Animals*.

67. For a different interpretation of how these reformers viewed purebred livestock, see Rosenberg, "No Scrubs," and Rosenberg, "The Trial of the Scrub Sire." Although Rosenberg rightly states that livestock reformers, many of whom were eugenicists, were attached to "antiquated racial logics," they did not, as he claims, see purebreds as fixed or permanent. In fact, these reformers unambiguously set out to reorient animals' bodies, including purebreds, to increase agricultural output. In this regard, they treated genotypes and phenotypes as complementary categories, considering phenotype to reflect body formations that correlated to market value and not just aesthetic traits like color.

68. Miles, *Stock-Breeding*, 2–10. See also Shaw, *Animal Breeding*. Shaw, expert in animal husbandry at the University of Minnesota, argues that livestock should be regarded as "machines for manufacturing agricultural products into forms more concentrated and possessed of a higher value" (5). He notes that making machines out of livestock geared toward surplus production led to greater profits for the farmer but that surplus agriculture required farmers to select livestock breeds that specialized in the production of a single commodity and to mate their animals with the goal of efficiency responsive to market demands in mind.

69. "Pure Bred Sheep"; Marshall, "A Study in Scotch Pedigree."

70. Plumb, "Purebred Live Stock Associations and Their Methods: Methods of Registering Live Stock" (1907), vol. 4, CSPP-OSU. See also Marshall, *Breeding Farm Animals*. Marshall argues that promotion was a central responsibility of a well-functioning breed association. Promotion came in different forms; many breed societies advertised for the breeders to encourage the sale of their animals over competing

breeds. Associations utilized livestock expositions to feature their breed and provided money for prizes and premiums to stimulate interest in raising and showing the breed, which increased their credibility and attracted the attention of the public, serving to educate it on the accomplishments and the value of their "superior stock."

71. *A Review of the International Live Stock Exposition*, 1918. The cattle department hosted breed shows for Shorthorns, Aberdeen-Angus, Herefords, Red Polls, Galloways, and Polled Durhams. The sheep department included shows for Shropshires, Hampshires, Oxfords, Lincolns, Cotswolds, Dorsets, Southdowns, Cheviots, Leicesters, and Rambouillets. The swine department offered purebred representation to Berkshires, Poland-Chinas, Duroc-Jerseys, Chester Whites, Hampshires, Yorkshires, and Tamworths.

72. Plumb, "Purebred Live Stock Associations and Their Methods: The Herd, Flock and Stud Book" (1907), vol. 4, CSPP-OSU. For additional information on the importance of pedigrees, animal name, and identification, see Marshall, *Breeding Farm Animals*, who argues that like humans, "well-bred" animals required names associated with their lineage to indicate both purity and quality, although he adds that in contrast to humans, breeders also used prominent female names to demonstrate the origination of high-quality genetics, physiological type, or production value. Often the "foundress," or stud female, was as important to a genetic line as the sire, and Marshall advises producers to keep the female name instead of the male for the female offspring of the foundress, particularly in the case of imported females because the imported female was especially prestigious. The breeder name also conveyed certain ideas about the animal related to quality, standards, and genetic families: for defamed breeders, the name carried negative value; for successful breeders, the name heightened the prestige of the animal. Each pedigree and certificate of registration required the breeder and owner, in case the breeder and owners were different people, to indicate to buyers and competitors the farm from which the animal came. See "Wool Investigations at the Ohio Experiment Station," 38–39.

73. *Review of the First International Live Stock Exposition.*

74. Browning, "A Short History of Ohio Chief 8727A." On the broader influence of Ohio Chief on the breed, see Evans, *History of the Duroc*. Evans founded the *Duroc Bulletin* and served as the secretary of the National Duroc Jersey Record Association and the American Swine Breeders' Association. For a discussion on the central role of the International and the "breeding up" goals of progressive agriculturalists in the hog industry, see "The International Swine Show." The author, frustrated with poorly bred hogs, complains about the ineffectiveness of state fairs in encouraging the use of purebred stock, but notes that the International deserved credit for normalizing the use of "improved genetics"; the International held breeders to high standards, and judges did not use superlatives to describe mediocre animals like at state fairs. The International, instead, institutionalized the importance of "prepotency" in sires and dams, like Ohio Chief, and the necessity of recording pedigrees.

75. Browning, "A Short History of Ohio Chief 8727A." In the *History of the Duroc*, Evans remarks that wherever the Duroc existed in the United States, the name Ohio Chief and the names of his progeny became widely recognized.

76. Genetic selection and reproduction of purebred animals limited the biological outcomes of food-producing animals—a eugenic trope that directly influenced the

curriculum of land-grant universities well into the second half of the twentieth century. Land-grant universities provided a unique home for eugenics by offering courses that tackled the question of the usefulness of eugenics. The International served as a hub for US Department of Agriculture officials, land-grant university professors, and members of the agricultural press who sought to disseminate a "master race" ideology they hoped would transform the nation's farms. For more on eugenics in these agricultural networks, see Glenna, Gollnick, and Jones, "Eugenic Opportunity Structures."

77. *A Review of the International Live Stock Exposition*, 1918.

78. Plumb, *A Study of Farm Animals* (41).

79. Plumb, *A Study of Farm Animals*, 38–47. Agricultural colleges and breed associations also urged producers to "grade up" with purebred genetics. The Iowa Experiment Station conducted research on the value of purebred sires in grading up dairy cattle, and in the first cross of "very inferior scrub cows in a section of country where the people had never used pure-bred sires" with purebred sires, the experiment station recorded a significant increase in milk production, while the second cross, resulted a 194 percent increase in milk production and 138 percent increase in butter fat. This evidence substantiated Plumb's claims: "Whatever merit we have in our herds and flocks to-day, we need not hesitate to say is due to the careful work of men who have used pure-breds only" (41).

80. De Loach, *Armour's Handbook of Agriculture*; De Loach, "Beef Cattle." See also "How to Increase Pork Supply," Sotham, "Building Meat on the Beef Model," and "Beef Cattle Breeding."

81. In a volume published by Armour's Bureau of Agricultural Research and Economics titled *Progressive Hog Raising*, Edward N. Wentworth defines the meaning of "breeding up" in the hog industry and the central role of purebred genetics. For Wentworth, "breeding up" and "grading up" are interchangeable terms. To breed up, a livestock producer first needed to understand the broader influence that purebred genetics had on herd uniformity and performance. As opposed to "unselected hogs" or pigs with unknown heritage, "selected," purebred hogs, Wentworth argues, transmitted physiological traits with more certainty and regularity. Wentworth believed that the fewer number of crossbred animals in an offspring's ancestry the more likely that offspring would carry in its body and genetic makeup the desired characteristics for meat production. Thus, with a transgenerational approach to the use of purebred hogs, progressive farmers would eliminate the "razorback" pig or the "inferior" hog, just as they had eliminated the scrub steer.

82. De Loach, *Armour's Handbook of Agriculture*; De Loach, "Beef Cattle."

83. "General Review"; Spoor, "Tells of Great Year," 11–12.

84. Burch, *Some Tested Methods for Livestock Improvement*; Mohler, "*Better Sires—Better Stock*"; "Better Sires—Better Stock"; "President Wilson Enrolls Flock in Better Sires Campaign"; "Steady Progress Shown in Campaign for Better Sires"; "Campaign for Better Sires Now Includes Nearly 400,000 Head of Stock Enrolled." Because scrubs were seen as enemies of the state, some farmers referred to them as Bolsheviks. See "A New Name for the Scrubs."

85. Burch, *Some Tested Methods for Livestock Improvement*; Mohler, "*Better Sires—Better Stock*"; "Better Sires—Better Stock"; "President Wilson Enrolls Flock in

Better Sires Campaign"; "Campaign for Better Sires Now Includes Nearly 400,000 Head of Stock Enrolled"; "Tendency Shown Toward Use of Purebred Female Stock"; Rosenberg, "No Scrubs"; Rosenberg, "The Trial of the Scrub Sire."

86. *Outline for Conducting a Scrub Sire Trial*; Burch, *Some Tested Methods for Livestock Improvement*; Rosenberg, "No Scrubs"; Rosenberg, "The Trial of the Scrub Sire"; "Finds Scrub Bull Guilty"; "Scrub Bull to Be Put on Trial!"

87. Burch, *Some Tested Methods for Livestock Improvement*; "Michigan for Better Dairy Stock"; "'Better Sires Demonstration Train' in Michigan"; "Dairy Demonstration Tour in Michigan."

Chapter 3 • *Recreating the Animal Body*

1. Fred Hartman's father bought his first Cheviots from Howard Keim in 1893, who established the breed in the Corn Belt after purchasing sixty-eight Cheviot rams and ewes in 1891 from flocks in Otsego County, New York, which contained nearly all the Cheviots in the United States imported from Britain. The breed considered Keim's farm the "pioneer flock of the west" (*Flock Book of the National Cheviot Sheep Society*, 13), and animals he produced, whether shown by him or customers, received acclaim at the Chicago's World Fair in 1893 and many state fairs with special recognition at the Illinois State Fair and Indiana State Fair. Keim's operation served as a stud farm for western Cheviot breeders and spawned many regional flocks, including Hartman's. In central Indiana alone, farmers had established eight Cheviot flocks by 1898.

2. Craig, *Judging Live Stock*. Craig was a professor of animal husbandry at the University of Wisconsin and Iowa Agricultural College, a dean at Texas Agricultural and Mechanical College, a director of the Oklahoma Agricultural Experiment Station; he also served as the editor of the *Canadian Live Stock Journal*. His contemporaries recognized Craig for creating the standards and process for livestock evaluation. In *Judging Live Stock*, he sets out the criteria and approach for judging horses, cattle, sheep, and hogs and describes the proper way for judges to handle livestock. To effectively evaluate sheep, judges should begin at the head, looking at the teeth to estimate age, the eyes to determine relative health, and the head to ensure that the sheep meets the aesthetic qualifications of the breed. Then the judge should feel the neck, brisket, and chest to assess depth and muscularity and look for straightness of top and structural correctness and also at the width, length, and depth of the most valuable market products: leg/rump, loin, and rib/rack. See also "Prof. John A. Craig," and "Death of Prof. John A. Craig."

3. *Flock Book of the National Cheviot Sheep Society*; "A Good Record"; "Hartman's Cheviots"; "Mr. F. B. Hartman"; "Maple Grove Flock Doing Well"; Hartman, "The Cheviot." Hartman did not attend a land-grant university, but as a young farmer he enthusiastically embraced purebred livestock and adopted progressive farming. He earned a reputation across the nation for being diligent and successful, and he was also known for a strong-headed desire for perfection in his sheep. Hartman put many of his sheep up for sale in 1904, after his father died.

4. Miles, *Stock-Breeding*, 7.

5. De Loach, *Armour's Handbook of Agriculture*.

6. "Charles Sumner Plumb," MSF 312, folder 6, CSPP-PU.

7. Plumb, "A Type of Breed."

8. Plumb, "A Type of Breed." See also Curtis, *The Fundamentals of Live Stock Judging and Selection*, 22–46. Curtis elaborates on Plumb's focused approach to breeding. The nature of the production goal dictated the type of animal to be raised; for Curtis, a "clear understanding of the purpose for which an animal is bred" and a "distinct conception of the type of animal adopted for the standard" are "necessary for convincing, uniform, clear-cut decisions" (46). At livestock exhibitions in particular, Curtis believed that "modern show ring judging [was] based" on the evaluation and utility of type and form; after identifying and sorting animals based on form, then the modern judge based his placings on quality within a certain type of livestock. Only using this modern evaluation could animal production on each farm conform to purpose—a prerequisite for Curtis in the mechanization of livestock.

9. Gay, *The Principles and Practice of Judging Live-Stock*, 3–23. Gay sees efficiency as a central goal for improved agriculture, and he defines what that meant for the farmer. He shows how the term "mechanical efficiency," one used to describe the manufacturing of nonanimal products, has meaning for the livestock industry and lists the attributes that need to be considered in aiming for such efficiency, such as the character of the materials of construction, the perfection of the constituent parts, accuracy of assembly, operational power, and the ability to control and effectively use the mechanical products. Gay correlates each of these production qualities with animal characteristics and encourages farmers to take up an elementary study of histology, anatomy, physiology, and pathology so they could link function and parts to the mechanization of the animal body.

10. Gay, *The Principles and Practice of Judging Live-Stock*, 4; *Lectures on the Results of the Exhibition Delivered Before the Society of Arts, Manufactures, and Commerce*, 59–98; Piggott, *Palace of the People*, 1–30.

11. Curtis, *The Fundamentals of Live Stock Judging and Selection*. See also Plumb, "Judging Stock at the Colleges. Plumb argues that collegiate judging was central to the education of future breeders, especially in the important task of correlating animal body type, or the exterior qualities of livestock, to the performance of the animal at the slaughterhouse.

12. Curtis, *The Fundamentals of Live Stock Judging and Selection*, 24–28. Curtis argues the displacement of "the original long-horn steer" by "the symmetrical, deep-set, well-developed, compact form of the modern bullock" over the course of the first two decades of the century marked considerable progress in the correlation between animal type and market function (17). This contention reflects two overlapping conversations among agriculturalists, namely, that advancing animal agriculture required a mechanical understanding of animal form and function and that for the constructive breeder, these mechanical animals assumed specific characteristics that necessitated the alteration of livestock breeding and animal body type. According to Curtis, the form most ideal for the market in meat-producing animals was a smaller-statured, compact animal with a square-like appearance.

13. Curtis, *The Fundamentals of Live Stock Judging and Selection*, 25.

14. Curtis, *The Fundamentals of Live Stock Judging and Selection*, 82–83.

15. Lloyd-Jones, "What is a Breed?" Mamie's Minnie specialized in milk production, not beef. This cow produced high rates of milk; she generated 14,838 pounds of milk in 1913 and 16,201 pounds in 1914. This picture was taken on the farm and used by

Lloyd-Jones to demonstrate the physiological differences in animal type (535). Mamie's Minnie possessed a prominent hipbone, which the ideal beef-producing cow did not have, because her calories were diverted from the production of meat and fat to milk.

16. *Review of the First International Live Stock Exposition*, 45–47.

17. Plumb, *A Study of Farm Animals*, 220–25; Marshall, *Breeding Farm Animals*, 209–21; Wentworth, *Progressive Beef Cattle Raising*; Allen, *American Cattle*, 45–61, 134–65; Plumb, "To Identify Breeds" (1922), vol. 4, CSPP-OSU.

18. Plumb, *Beginnings in Animal Husbandry*, 77–90; Plumb, *A Study of Farm Animals*, 338–47; De Loach and Phillips, *Progressive Sheep Raising*; Stewart, *The Domestic Sheep*, 22–101. Stewart argues that English sheep outperformed sheep from other countries, especially France. Even at French agricultural exhibits, imported British sheep dominated continental sheep in the physiological traits necessary for meat production. Accordingly, Stewart declares, American shepherds had little to learn from French sheep breeders except "to discover the effects of ages of neglect" (52). Cheviots produced food and fiber, as did all British sheep breeds; however, meat production in the breed greatly outpaced the value garnered from wool. Hailing from the Cheviot Hills on the border of England and Scotland, this hornless breed was hardy and durable as well as, as Plumb observed, one of the prettiest. With white hair cover on the head and legs and a black nose and black hooves, when prepared for the show ring, the Cheviot was striking with a regal, up-headed stature. Rams and ewes reached maturity at 200 pounds and 150 pounds, respectively. Southdowns, in contrast, possessed a wool-covered face of reddish-brown color. They had short heads and short necks, and they built a reputation in the United Stated for being a breed chiefly for mutton. Among sheep breeds, Southdowns matured the earliest and had short, blocky bodies with a thick, meaty leg. Admired by butchers, Southdowns yielded a high percentage of meat with little waste, and even though they performed poorly on the range, they met the needs of the modern meatpacking industry on the pastures and feedlots of the Corn Belt.

19. Henry and Morrison, *Feeds and Feeding Abridged*, 296–376.

20. Coburn, *Swine Husbandry*, 21–80; Plumb, *Beginnings in Animal Husbandry*, 99–113; Plumb, *A Study of Farm Animals*, 389–403; Wentworth, *Progressive Hog Raising*; Shaw, *The Study of Breeds in America*, 276–300; Plumb, "To Identify Breeds" (1923), vol. 4, CSPP-OSU. As a foundational breed for the Poland China, the Berkshire went through a period of great popularity from 1831 to 1841, a fad referred to "the Berkshire fever." As a result of decreasing interest midcentury, breeders neglected Berkshires and made little improvements. Following the American Civil War, breeders began to import Berkshires from England again because of their muscularity, efficient feed conversion rate, fertility and prolific production of offspring, and uniformity in color and quality. Not unlike the Poland China, the Berkshire had a black body with white points, but their body shapes distinguished the breeds. Berkshires were not as coarse or as compact as Poland Chinas and generally had a longer, more angular appearance. Mature Berkshire boars weighed five hundred pounds and mature sows four hundred pounds. Yorkshires offered a direct contrast to the body type of Berkshires and Poland Chinas. Often referred to as Large Yorkshires, this breed fell into the bacon category. Originally from England, the Yorkshire came from the oldest line of breeding. Yorkshires had large frames with a narrow body. The head, which was longer than in the

lard types, inclined forward with erect ears. The body of the Yorkshire was long and deep and had smooth sides—an advantage in bacon production. Other popular breeds in the United States included Chester Whites, Duroc-Jerseys, Hampshires, Tamworths, Cheshires, Victorias, and the Essex.

21. Irwin, "Agricultural Events at the 1904 St. Louis World's Fair." For more on the allure of monstrous or freak animals and plants, see Pawley, *The Nature of the Future*.

22. Pawley, *The Nature of the Future*, 1–19.

23. Sanders, *The Story of the International Live Stock Exposition from its inception in 1900 to the Show of 1941*, 8–9; Poole, "The Twentieth International." Poole was a renowned expert on the Chicago livestock market and served as a journalist for *The Breeder's Gazette*. At the end of the nineteenth century, Poole remembers, the amount of fat and overwhelming size of the show steers presented obstacles to quality and efficiency on the farm. Showmen exhibited steers that weighed over a ton at old fat stock shows, and the first International consisted of, Poole notes, "aged steers freighted with fat" (14). Both overly conditioned steers and aged steers limited agricultural productivity and the ability of farmers to increase the quality of marketed products and their potential earnings.

24. Ritvo, *The Animal Estate*, 75.

25. Ritvo, *The Animal Estate*. See also Pawley, *The Nature of the Future*. Pawley describes the role of "agricultural monsters" in validating or demonstrating human influence over biological beings, and as a result, human accomplishment, which often reflected settler colonialism dogmas demonstrating "white ingenuity" associated with agriculture, empire, and expansion (2, 4).

26. Stewart, *Feeding Animals*, 528–34; Olmstead and Rhode, *Creating Abundance*, 271–73.

27. In "Market Classes and Grades of Cattle with Suggestions for Interpreting Market Quotations," Herbert W. Mumford outlines the different types of cattle and categorizes them by quality and desirability, highlighting the shift in animal husbandry toward young, more efficient market livestock. Mumford includes the weights and ages necessary to minimize cost and maximize value. Even though Mumford argues that weight was less of a concern than quality and condition, the decrease in finish weight had a compounding impact on quality and condition; thus, there was a strong correlation between weight and quality.

28. Plumb, "Big Type," vol. 4, CSPP-OSU. Not only did this trend directly impact the cattle industry, but sheep and hogs also dramatically changed. Thus, in all meat-producing livestock, the International provided a mechanism to orient breeding toward the "profitable type." For example, feeding a hog for eighteen months to the market weight of 615 pounds was unprofitable despite the larger size. Feeding a hog to 220 pounds in five months, by contrast, was profitable; the extra days of feeding not only cost more, but the additional age and weight eventually decreased the value of the carcass as well.

29. Poole, "The Twentieth International."

30. *A Review of the International Live Stock Exposition*, 1918, 22–23.

31. Brown, "From the International Judge."

32. Benson, "The Mental Processes of a Stock Judge." Judges came under scrutiny and criticism in the show ring just like the animals they evaluated. The judges

influenced breeders with their perspective, and, their function went beyond mediating or arbitrating shows like a referee. They had tastes, preferences, and priorities that might differ from those of exhibitors or observers. Often onlookers and farmers disagreed with the opinion of the judge, but for judges to maintain credibility, promoters of progressive agriculture argued that they should uphold, in a strict and open way, their honest opinions, which required them to avoid favoritism or bias.

33. *Review of the First International Live Stock Exposition*, 55.

34. *A Review of the International Live Stock Exposition*, 1916, 85.

35. Benson, "The Mental Processes of a Stock Judge," 81; Plumb, "Ohio Livestock Improvement" (1922), vol. 4, CSPP-OSU; Plumb, "Feeding Baby Beef" (1915), vol. 4, CSPP-OSU.

36. Curtis, *The Fundamentals of Live Stock Judging and Selection*, 43–45.

37. Gay, *The Principles and Practice of Judging Live-Stock*, 58–59.

38. Curtis, *The Fundamentals of Live Stock Judging and Selection*, 40–42.

39. Gay, *The Principles and Practice of Judging Live-Stock*, 64–65.

40. Benson, "The Mental Processes of a Stock Judge," 81.

41. Benson, "The Mental Processes of a Stock Judge," 81; Curtis, *The Fundamentals of Live Stock Judging and Selection*, 40–42; Gay, *The Principles and Practice of Judging Live-Stock*, 86–90.

42. *Review of the First International Live Stock Exposition*, 47.

43. The photograph of Goldie's Ruby for the 1918 exposition showed her as deep bodied, with her brisket, chest, and stomach barely clearing the straw bedding. The photograph itself demonstrated the goals of the International. By eliminating the space between her underside and the ground, almost appearing to have no legs at all, the photographer was emphasizing the ideal body type of that era.

44. *A Review of the International Live Stock Exposition*, 1918, 27; Sanders, "The Golden Age of Shorthorns."

45. Sanders, "The Golden Age," 10.

46. Sanders, "The Golden Age," 10.

47. Benson, "The Mental Processes of a Stock Judge," 81. See also Sanders, "Are You Doing Your Part?" (1916), box 4, file 32, AHSP. The religious language related to spreading ideas or changing beliefs permeated the discourse. However, people were not only the disciples of modern agriculture; animals were too. In "Are You Doing Your Part?" Sanders details the advantages of improved livestock, highlighting the role of Shorthorns in revolutionizing the agricultural community wherever they were bought, bred, and raised. Shorthorns as a breed served as missionaries of "modern" agriculture and were assigned the solemn responsibility of displacing the "heathen" Longhorns from the ranges in the west. Shorthorns, Sanders maintains, certainly would improve the "bovine heathens of the earth" on typical "Mexican haciendas," on "Australian stations, and "African veldts," and on what was without a doubt the "richest large agricultural area in the world—the American corn belt." He notes how ranchers on one range in El Paso raised a fourteen-month-old Shorthorn steer with Longhorns and how at that young age, the Shorthorn weighed twenty-five to fifty pounds more than the Longhorns, which were all between five to ten years old, and was worth $150 more. For Sanders, raising these improved animals, especially Shorthorn cattle, offered the breeder additional revenue and also provided a service—of the missionary sort—to the public.

48. Plumb, "Teaching Animal Selection."

49. Plumb, "Teaching Animal Selection."

50. Plumb, "Teaching Animal Selection."

51. Plumb, *A Study of Farm Animals.*

52. Curtis, *The Fundamentals of Live Stock Judging and Selection*, 22–46.

53. Curtis, *The Fundamentals of Live Stock Judging and Selection*, 222–73.

54. Plumb, *A Study of Farm Animals.*

55. Gay, *The Principles and Practice of Judging Live-Stock*, 239–67. According to Gay, a good sheep's meat yield could be between 45 and 63 percent.

56. Curtis, *The Fundamentals of Live Stock Judging and Selection*, 51–57, 366–428. To evaluate wool on a live animal, the breeder or judge parted the fleece with their hands. In general, there were three types of wool: fine wools, medium wools, and long wools. The majority of mutton breeds were medium wools. Regardless of wool type, the judge parted the fleece to examine the staple, but with the different wool types the judge evaluated them based on their unique characteristics. Regardless of type, judges looked for uniformity and consistency of wool across the entire body. This demonstrated a higher-yielding and higher-quality fleece. A judge had to balance quality, quantity, and uniformity to gauge the overall rating of the fleece produced by both wool and mutton sheep.

57. Plumb, *A Study of Farm Animals.*

58. Steers, wethers, and barrows were castrated cattle, sheep, and hogs, respectively. In the cattle category, steers also competed against spayed or martin heifers. Martin (freemartin) heifers were sterile females with masculine features that resulted from being born as a twin of a bull calf.

59. *A Review of the International Live Stock Exposition*, 1917, 88, 150–54, 182–86.

60. *Review of the First International Live Stock Exposition*, 51.

61. *Review of the First International Live Stock Exposition*, 150–52, 165.

62. *Review of the First International Live Stock Exposition*, 7–9; Plumb, "International Live Stock Exposition" (1900), vol. 3, CSPP-OSU.

63. *Review of the First International Live Stock Exposition*, 1900, 7–9; Plumb, "International Live Stock Exposition" (1900), vol. 3, CSPP-OSU.

Chapter 4 • New Animals, New Problems

1. "The Little International," Communal Accessions 6, 16D1, folder 10, CADAS; Whitford, *For the Good of the Farmer*, 265–67. Purdue's Department of Animal Husbandry hosted the event with the undergraduate-run Hoof and Horn Club.

2. "The Little International," Communal Accessions 6, 16D1, folder 10, CADAS; Whitford, *For the Good of the Farmer*, 265–66. Also see "Program: Little International Livestock Show and Fitting Contest, 1925," box 1, SASC, and "Fourth Annual Little International Livestock Show and Fitting Contest, 1926," box 1, SASC. The Saddle and Sirloin Club at North Dakota Agricultural College launched Fargo's version of the Little International shortly after World War I ended. Similar to Purdue, North Dakota Agricultural College highlighted the importance of this campus event in training students and livestock for the most important show of the year—the International. Students at North Dakota Agricultural College provided musical and theatrical shows in between the showing and placing of cattle, sheep, pigs, horses, and poultry. To

entertain crowds in 1926, the school's Gold Star Band played, and the Alpha Gamma Rho Quartette sang, and a small cast of students performed "The Little Red Mare," a one-act play.

3. Gobble, "Indiana and Purdue Win Again."

4. "The Little International," Communal Accessions 6, 16D1, folder 10, CADAS; Whitford, *For the Good of the Farmer*, 274–76.

5. *Review of the First International Live Stock Exposition*, 12–13, 156.

6. Curtiss, "Brief History of the American Society of Animal Production," box 1, folder 2, PSSP; Curtiss, "Some Foundations in Agricultural Education," MS LB2543 C947s, Iowa State University Library, Ames.

7. Campbell, "He Put the 'A' in Agriculture," box 1, folder 8, CFCP; Beckman, "Dean C. F. Curtiss," box 1, folder 8, CFCP; Randolph, "C. F. Curtiss Dies," box 1, folder 1, CFCP; Pammel to Coolidge (1924), box 1, folder 8, CFCP.

8. Curtiss, "Some Foundations in Agricultural Education," MS LB2543 C947s, Iowa State University Library, Ames; Plumb, "Who Is the Scientific Farmer?" (1915), vol. 4, CSPP-OSU.

9. Davenport, "Scientific Farming," 45–50.

10. Veit, *Modern Food, Moral Food*, 101–11.

11. Veit, *Modern Food, Moral Food*, 101–11.

12. Christie, "The New Agriculture," (1916), MSF 89, folder 2, GICP; Christie, "Agricultural Extension Work," MSF 89, folder 2, GICP.

13. *Review of the First International Live Stock Exposition*, 165–66.

14. See Craig, *Judging Live Stock*.

15. Shepperd, *Livestock Judging Contests*.

16. Shepperd, *Livestock Judging Contests*.

17. Vaughan, *Types and Market Classes of Livestock*.

18. Shepperd, *Livestock Judging Contests*. Shepperd graduated from Iowa Agricultural College in 1891. At the time he attended the school, livestock judging classes did not exist, and the school provided little instruction in animal selection. "Scientific agriculture was at a low ebb" (5) among students, he notes, and he had the distinction of being the only agricultural student at the university. He went on to the University of Minnesota for graduate study, where he found similar conditions. Then, he moved to the University of Wisconsin, where J. A. Craig had developed a course in stock evaluation. Craig mimeographed a scorecard prototype for students to use while evaluating livestock, which he revised repeatedly until it was in publishable form. Under Craig's direction, Wisconsin became a foundational school for collegiate judging and courses on animal selection. In 1892, the University of Wisconsin held the first student judging contest on its campus—the Wisconsin Winter Short-Course Judging Contest. Following this event, collegiate animal selection became a popular course and activity in Madison and at other land-grant institutions. Colleges enhanced the short-course instructional format and subsequently developed two- and four-year collegiate judging programs.

19. Vaughan, "A Picture of the Live Stock Industry."

20. Vaughan, *Types and Market Classes of Livestock*; Shepperd, *Livestock Judging Contests*. See also Mumford, "Market Classes and Grades of Cattle with Suggestions for Interpreting Market Quotations," Dietrich, "Market Classes and Grades of Swine,"

Obrecht, "Market Classes and Grades of Horses and Mules," Coffey, "Market Classes and Grades of Sheep," and Hall, "Market Classes and Grades of Meat."

21. Plumb, "Students' Judging Contest." The format of five animals per class and five students per team changed over time to four animals and four students, as it is in current competitions, and eventually reasons were presented orally rather than on paper.

22. Plumb, "Students' Judging Contest."

23. Shepperd, *Livestock Judging Contests*. Despite the changes in format, the 1904 contest also was heavily criticized. The professional judges and International superintendents, although competent in animal selection and management, became the targets of negative feedback revolving around their limited ability to work and communicate with college professors and collegiate competitors.

24. Shepperd's determination and focus on improvement had an immediate impact on the protocol for the annual event—a format copied and mimicked by other events for years to come. The judging contest, consequently, attracted more participants and gained in popularity and prestige. However, Shepperd quickly learned that criticism, justified and not, would be an ongoing problem regardless of the precautions he took. In the 1920s, Shepperd received letters from coaches informing him of impropriety at other contests and warning him of potential problems at the International. J. H. Skinner, who was at Purdue University, wrote to Shepperd on November 19, 1924, to inform him of alleged cheating among college students and coaches. Skinner pushed Shepperd to take action to ensure that the coaches did not influence or cheat to gain an advantage, arguing the alteration or expansion of existing rules would help limit "the appearance of unfairness." Skinner suggested that preventing coaches from selecting livestock for the contest would help mitigate the problem and that measured distance, codified by rules, should be maintained between groups of contestants so that students from the same institution would be kept separate; these formalized barriers, Skinner hoped, would reduce or eliminate information sharing among teammates (box 14, JHSP). In a follow-up letter on November 26, 1925, Skinner went beyond offering helpful suggestions to alleging that coaches were plotting grand schemes to communicate placings to students. Coaches, many of whom had military training, could mark or label through code, Skinner argued, to communicate the class order to his students (box 14, JHSP). Two years later, Shepperd actively pursued a case of cheating at the International against the Oklahoma Agricultural and Mechanical College. In response, the president of the school, Bradford Knapp, said that the student in question would be immediately expelled if the investigation unveiled a legitimate attempt to cheat at the International. However, Knapp also noted that there had been many attempts to undermine his school because of its recent successes. These false allegations, he contended, amounted to attempts by rival institutions to defame the program (box 14, JHSP).

25. Shepperd, *Livestock Judging Contests*, 14–15; *Official Catalogue*, 373–77. Students received points for the placing of each class and the reasons for their rankings, which gave rise to two problems. First, many university professors did not trust practical agriculturalists to grade written reasons. To solve this problem, the committee assigned a college-trained expert to partner with the practical officials to grade the written portion. Second, how much credit organizers should give students who placed a class

correctly and provided poor reasons in comparison to the competitors who placed the class poorly but supplied good reasons was not a straightforward matter. In response, they established a set of standards and a point system to evaluate each separately and then add the results. See Plumb, "Students' Judging Contest," 152–53, and Plumb, "Students' Judging Contests Again."

26. Shepperd, *Livestock Judging Contests*, 20.

27. Shepperd, *Livestock Judging Contests*, 8–9.

28. Shepperd, *Livestock Judging Contests*, 10–11.

29. Although because of increased labor needs, women participated more in industry during World War I, by 1920, they made up a smaller percentage of the labor force than they did in 1910. See Kennedy, *Over Here*, O'Neill, *Everyone Was Brave*, and Chafe, *The American Woman*. According to these authors, the war and suffrage were highpoints for women in public life, and following the war, the spheres assigned to men and women that had existed before the war re-formed, and former suffragists found themselves divided by differences of opinions on a variety of political and cultural issues. Kennedy uses the Sheppard-Towner Act of 1921 to demonstrate this shift. The measure provided federally financed instruction for maternal and infant care and was supported by feminist groups, but, as Kennedy notes, it was not intended to encourage women to enter industry but to push them back into the prewar domestic sphere. For an account of labor force demands during the war, especially in the "battle of materials," see Hobsbawm, *The Age of Extremes*.

30. "Women as Animal Husbandmen." Women played a major role in political activities from abolition and suffrage to community reform and political demonstration. In fact, before the United States entered World War I, Fanny Garrison Villard, Carrie Chapman Catt, and Jane Addams, along with many others, formed the Woman's Peace Party. This organization, joined by the League to Enforce Peace and the American League to Limit Armaments, attempted to prevent an American military buildup and to protect the gains made by women in the political sphere. These organizations feared that the masculine and virile overtones of war could undo the progress that the feminist movement had made. For more on these antipreparedness organizations, see Kennedy, *Over Here*. For a broader discussion of the diversity, complexity, and vocabulary of feminism and political activism, see Cott, *The Grounding of Modern Feminism*.

31. *A Review of the International Live Stock Exposition*, 1918, 258.

32. Ashton, "Animal Husbandry."

33. Ashton, "Animal Husbandry." Ashton's publication also challenged eugenicists at the American Breeders' Association who argued that race improvement conflicted with the growing role of women in society. These reactionary observers opposed the extension of the vote to women, criticized women's activism in politics, and in particular worried about the impact of increased independence of women on the perpetuation of America's "best racial blood," arguing that women of "best blood" had a "racial responsibility" to reproduce. See "The Woman Movement and Eugenics."

34. Ashton, "Animal Husbandry," 13–14. After describing the historical and contemporary importance of women in agriculture, Ashton reminds readers of the educational and career-related limitations women faced in the field. She argues that women had skills that were useful to agricultural research and teaching at the

secondary and postsecondary levels because they taught many of the nation's children in traditional disciplines like Latin and so that with the expansion of agricultural education under the Smith-Hughes Act of 1917, they should be agricultural educators as well.

35. "Deaths."

36. Ashton, "Animal Husbandry," 13–14.

37. Whitford, *For the Good of the Farmer*, 274–76; McCartney, "Merry Monarch a Great Steer"; Gobble, "Indiana and Purdue Win Again."

38. Whitford, *For the Good of the Farmer*, 276.

39. Whitford, *For the Good of the Farmer*, 272.

40. In *The 4-H Harvest*, Rosenberg argues that 4-H simultaneously normalized systems of agriculture—capital-intensive crop and animal husbandry regimes, in particular—and "the gendered production of desirable bodies through heteronormative family farms" (10). This intersection of farm production and the state control of gender and sexuality manifested in 4-H education; "proper" gendered labor and farm management tied together American nationalism, civic duty, and white commercial family farms. The International, like 4-H, worked to normalize behavior and preferences and tastes for the type of cow, sheep, or hog suited for the modern farm.

41. "Management of Purdue's Angus Herd," Communal Accessions 6, 16D1, folder 10, CADAS; "The Little International," Communal Accessions 6, 16D1, folder 10, CADAS.

42. "Management of Purdue's Angus Herd," Communal Accessions; "The Little International," Communal Accessions 6, 16D1, folder 10, CADAS; Whitford, *For the Good of the Farmer*, 279.

43. Animal nutrition researchers met at Cornell University on July 28, 1908, to initiate the organization of this association. H. P. Armsby, dean of the School of Agriculture at Pennsylvania State College and the leading animal nutritionist in the United States, headed up the group. W. H. Jordan, director of the New York State Agricultural Experiment Station, H. J. Waters, director of the University of the State of Missouri Agricultural Experiment Station, H. R. Smith, animal husbandry expert at the Nebraska Agricultural Experiment Station, and J. H. Skinner, dean of the School of Agriculture at Purdue University, also attended this first meeting. Representatives from the US Department of Agriculture and thirteen experiment stations also met with Armsby.

44. Armsby, "The Food Supply of the Future."

45. "Report of the Committee on Terminology," (122).

46. Stewart, *The Domestic Sheep*; Clarke, *Fitting Sheep for the Show Ring and Market*, 75–79; De Loach and Phillips, *Progressive Sheep Raising*; Wentworth, *Progressive Hog Raising*; Wentworth, *Progressive Beef Cattle*; Smith, *Profitable Stock Feeding*; Henry and Morrison, *Feeds and Feeding*; Coburn, *Swine Husbandry*; Mumford, *Beef Production*.

47. *A Review of the International Live Stock Exposition*, 1913; *A Review of the International Live Stock Exposition*, 1916; *A Review of the International Live Stock Exposition*, 1917; *A Review of the International Live Stock Exposition*, 1921; *A Review of the International Live Stock Exposition*, 1922.

48. Land-grant universities and associated public-funded agricultural institutions more formally intervened in American agriculture after the passage of the Adams Act

in 1906, which financially aided experiment stations in their research, and after the passage of Smith-Lever Act in 1914, which led to the creation of a nationwide extension service that connected the government and researchers to farmers. See Ferleger, "Arming American Agriculture for the Twentieth Century."

49. Curtiss, "Some Foundations in Agricultural Education."

50. "Food Production and Conservation Committee": 10–11; "Food Program for Illinois"; Crissey, "First Aids to Farmers."

51. "Banks Form Syndicate to Supply Seed Corn"; "Food Program for Illinois"; Crissey, "First Aids to Farmers"; *A Review of the International Live Stock Exposition*, 1918, 262; *A Review of the International Live Stock Exposition*, 1919, 271–72.

52. Hackleman, *History*, 7–9, 9–10, 10–11; "Food Production and Conservation Committee"; "Banks Form Syndicate to Supply Seed Corn"; "Food Program for Illinois"; "First Aids to Farmers"; *A Review of the International Live Stock Exposition*, 1919, 271–78.

53. *A Review of the International Live Stock Exposition*, 1922, 283; *A Review of the International Live Stock Exposition*, 1921, 299; Rogers, "Purdue's Experimental Farms"; "Indiana Leads in Winnings"; "The Fourth International Grain and Hay Show."

54. De Loach and Phillips, *Progressive Sheep Raising*; Wentworth, *Progressive Hog Raising*; Wentworth, *Progressive Beef Cattle Raising*; Stewart, *The Domestic Sheep*; Clarke, *Fitting Sheep*; Smith, *Profitable Stock Feeding*; Henry and Morrison, *Feeds and Feeding*; Coburn, *Swine Husbandry*; Mumford, *Beef Production*.

55. Sheets and Kelley, "Beef-Cattle Barns"; McWorter, "Equipment for Farm Sheep Raising"; De Loach and Phillips, *Progressive Sheep Raising*; Wentworth, *Progressive Hog Raising*; Wentworth, *Progressive Beef Cattle Raising*; Stewart, *The Domestic Sheep*; Clarke, *Fitting Sheep*; Smith, *Profitable Stock Feeding*; Henry and Morrison, *Feeds and Feeding*; Coburn, *Swine Husbandry*; Mumford, *Beef Production*.

56. *Plans of Farm Buildings for Western States*.

57. Sheets and Kelley, "Beef-Cattle Barns"; *Plans of Farm Buildings for Western States*; McWorter, "Equipment for Farm Sheep Raising"; De Loach and Phillips, *Progressive Sheep Raising*; Wentworth, *Progressive Hog Raising*; Wentworth, *Progressive Beef Cattle Raising*; Stewart, *The Domestic Sheep*; Clarke, *Fitting Sheep*; Smith, *Profitable Stock Feeding*; Henry and Morrison, *Feeds and Feeding*; Coburn, *Swine Husbandry*; Mumford, *Beef Production*. Architecturally, late twentieth-century hog-feeding barns differed from the earlier confinement buildings, but the emphasis on high-population fattening programs already prevailed among reformers and progressive feeders.

58. Mumford, *Beef Production*, 145–49.

59. Fitzgerald, *The Business of Breeding*, 4.

Chapter 5 • New Animals, New Problems

1. Toole, "Development of Our Modern Beef Type."

2. Poole, "The Twentieth International"; Tormey, "International Just Out of Its Teens"; "20th Anniversary of International"; Poole, "The International Anniversary Show"; "The World's Greatest Stock Show"; "A Big Reunion"; "International Livestock Exposition"; "The International Livestock Show."

3. Poole, "The Twentieth International"; "Has the International Made Good?"; "Experts Tell Radio World Romantic Story of Live Stock"; "'Jim' Poole Says the International Has Missed Its Calling."

4. Chandler, "Breeding for a Purpose"; Poole, "Young Cattle the Most Profitable"; "Has the International Made Good?"; "'Jim' Poole Says the International Has Missed Its Calling"; Olmstead and Rhode, *Creating Abundance*, 314–29; "The Production of Baby Beef"; Alexander and Alexander, "Farming"; Whitson, "Baby Beefmaking in the Cornbelt"; Krueck, "Topping the Market with Baby Beef"; Beresford, "Baby Beef Making Safest System for Corn Belt"; Toole, "Development of Our Modern Beef Type."

5. Poole, "The Twentieth International"; "Has the International Made Good?"; "Experts Tell Radio World Romantic Story of Live Stock"; "'Jim' Poole Says the International Has Missed Its Calling."

6. "The International Answers the Nation's Call"; "The Great Livestock Exposition"; "The 1917 'International'"; *A Review of the International Live Stock Exposition, 1917*; *A Review of the International Live Stock Exposition, 1918*.

7. "The International Answers the Nation's Call"; "The Great Livestock Exposition"; "The 1917 'International'"; *A Review of the International Live Stock Exposition, 1917*; *A Review of the International Live Stock Exposition, 1918*.

8. Poole, "The Twentieth International"; "Has the International Made Good?"; "Experts Tell Radio World Romantic Story of Live Stock"; "'Jim' Poole Says the International Has Missed Its Calling."

9. Poole, "The Twentieth International"; "Has the International Made Good?"; "Experts Tell Radio World Romantic Story of Live Stock"; "'Jim' Poole Says the International Has Missed Its Calling"; "'Better Sires' Campaign Started by 'Uncle Sam.'" The Aberdeen-Angus Breeders' Association journal featured a column accusing Poole of shifting his market analysis to suit the opinions of his readers. When writing for the *Breeder's Gazette*, for example, he applauded progressive goals and lauded the International and when he published in the *Producer*, he said what the ranchers in the West wanted to hear. The Aberdeen-Angus breeders believed that this oscillation of opinion undermined the force of his criticism.

10. *Yearbook of the United States Department of Agriculture, 1921*, 730–31; Schlebecker, *Cattle Raising on the Plains*, 84; "Remarkable Classes of Beef Breeding Cattle"; Minster, "Hereford Claims from England"; Poole, "The Twentieth International"; "Has the International Made Good?"; "Experts Tell Radio World Romantic Story of Live Stock"; "'Jim' Poole Says the International Has Missed Its Calling"; "'Better Sires' Campaign Started by 'Uncle Sam.'"

11. Sotham, "The Potency of Hereford Blood."

12. Pickett, "White Faces"; Schlebecker, *Cattle Raising on the Plains*, 84; McGavock, "The Trend of the Times in the American Hereford Trade"; Minster, "Hereford Claims from England."

13. Warwick, "Fifty Years of Progress in Breeding Beef Cattle"; Terrill, "Fifty Years of Progress in Sheep Breeding."

14. "Baby Beef" (1904); Poole, "The Twentieth International"; "Has the International Made Good?"; "Experts Tell Radio World Romantic Story of Live Stock"; Wing, "The Present Situation in the Live Stock World"; "'Jim' Poole Says the International

Has Missed Its Calling"; "Our Beef Supply"; Darlow, "Fifty Years of Livestock Judging"; Willham, "Genetic Improvement of Beef Cattle in the United States."

15. Algie Martin Simons, editor of the *International Socialist Review*, objected to what he saw as the theft of money and resources from the countryside by urban consumers and capitalists. He lamented that the owners of agricultural mortgages and landlords far too often resided in the city and rented property to tenant farmers, creating not just a physical distance between the owners of the farm and the work and the workers of the farm but also an emotional one. Tenant farmers shipped the products of the soil's "choicest elements," fertility, to the city, and that essential organic matter was used in the urban by-product markets and led to pollution instead of being returned to the field, and the waste filled the rivers and lakes with toxic material. Simons also objected to the impact of modern life and modern farming on producers. He urged people to reject the imposition of crop and animal husbandry systems geared toward surplus production. He especially shunned the idea that farmers had to be specialists and forced into modern farming, as that rendered them dependent on a network of other specialists, salespeople, outside inputs, and capital. See his *The American Farmer*.

16. Fitzgerald, *Every Farm a Factory*, 15.

17. Hopkins, *Soil Fertility and Permanent Agriculture*, xvii–xxiii, 198–235.

18. Hopkins, *Soil Fertility and Permanent Agriculture*, 198–235; "Facts about Soil Fertility"; Bryant, "Raising Bumper Grain Crops."

19. Smil, *Enriching the Earth*; Hopkins, *Soil Fertility and Permanent Agriculture*, 198–235; "Facts about Soil Fertility"; Bryant, "Raising Bumper Grain Crops."

20. Jordan, *North American Cattle-Ranching Frontiers*, x, 269–272; Olmstead and Rhode, *Creating Abundance*, 264.

21. Olmstead and Rhode, *Creating Abundance*, 314–29; Morrison, *Feeds and Feeding*, 405–6; Stewart, *Feeding Animals*.

22. *A Review of the International Live Stock Exposition*, 1916, 240–43; *A Review of the International Live Stock Exposition*, 1918, 252–55; *A Review of the International Live Stock Exposition*, 1922, 278–317.

23. "A Productive Stock Farm."

24. Coburn, *Swine Husbandry*, 140–49; Wentworth, *Progressive Hog Raising*; "Concrete for the Farm." Many hog producers included self-feeders on these concrete slabs so that the pigs could eat as much grain as possible. At the base, the feeders had doors or flaps that covered holes. Gravity kept the holes filled with grain and the hogs lifted the doors with their snouts to access the feed. Simpler in design, these feeders allowed producers to give hogs a maximum quantity of grain-based calories. Other hog producers invested in large barns with narrow but long pens with troughs and waterers. This controlled environment allowed producers to measure and dictate feed intake and weight gain.

25. "Vaccine Protects Feeder Lambs"; "Watch Your Feeder Lambs, They May Eat Themselves to Death"; "Entrerotoxemia Can Hit Healthy Lambs"; "Overeating Disease Is Serious Threat to Feeder Lambs."

26. Parasites also often caused anemia, and cattle and sheep fell victim to a particularly brutal type of parasite called coccidiosis that sucked nutrients from the

host. Their appetites decreased, their backs arched, and their bodies became weak and thin. Coccidiosis often led to pneumonia and even death. See "Permanent Buildings," "Vet Advises Treating Lambs for Worms," "Good Management Protects Calves from Coccidiosis," 4; "Paved Feedlots Found Profitable," and "Livestock Farming Conserves Soil."

27. Vaughan, *Types and Market Classes of Live Stock*, 69–78; "About Baby Beef"; "What Is Baby Beef?"; Ray, *The Production of Baby Beef*.

28. Poole, "Why Beef Consumption Is Lagging"; "Baby Beef"(1908).

29. Hayes, "Baby Beef Production"; Goodwin, "The Performance Record of Angus Cattle"; "The Carlots of Fat Cattle"; "Baby Beef," (1908); "About Baby Beef."

30. Poole, "Craze for Light Cattle"; "The Show of Carlots of Fat and Feeder Cattle"; Alexander and Alexander, "Farming"; Toole, "Development of Our Modern Beef Type."

31. *The Union Stock Yard and Transit Company of Chicago: Eighty-Second Annual Live Stock Report, Year 1947 and Summary for Years 1865 to 1947* (1948), box 4-93, CSC.

32. See "Average Weight of Livestock" for the years 1901 and 1912, Evvard, "Producing Baby Beeves on Corn Belt Farms," Beresford, "Baby Beef Making Safest System for Corn Belt," McGavock, "The Trend of the Times in the American Hereford Trade," and Ray, *The Production of Baby Beef*.

33. Curtis, *The Fundamentals of Live Stock Judging and Selection*, 40–42; Gay, *The Principles and Practice of Judging Live-Stock*, 64–65, 86–90; Benson, "The Mental Processes of a Stock Judge."

34. Black, "Beef and Dual-Purpose Cattle Breeding"; "Livestock Breeding at the Crossroads."

35. *A Review of the International Live Stock Exposition*, 1948, 59–75; Burke, Schaff, and Haag, *Central Illinois Aberdeen Angus Association Preview Show*); "Weber, Arthur D."

36. Warwick, "Fifty Years of Progress in Breeding Beef Cattle"; Willham, "Genetic Improvement of Beef Cattle in the United States."

37. Sanders, "When the Show-Ring Hurts the Breeds"; Maxwell, "An Analysis of the Results of the Steer Carcass Contest at the International Livestock Exposition."

38. "Points on Treatment of Herd Boars"; Myers, "Handling the Sows during the Breeding Season"; Boyce, "Breeding Crate Saves Time and Trouble"; "Good Enough for the Breeder and Cheap Enough for the Farmer"; "Smith's Standard Breeding Crate."

39. "Another Breeding Crate"; "Points on Treatment of Herd Boars"; Myers, "Handling the Sows During the Breeding Season"; Boyce, "Breeding Crate Saves Time and Trouble"; "Good Enough for the Breeder and Cheap Enough for the Farmer"; "Smith's Standard Breeding Crate."

40. Ritchie, "From Big to Small to Big to Small," pts. 2 and 3; Warwick, "Fifty Years of Progress in Breeding Beef Cattle"; Willham, "Genetic Improvement of Beef Cattle in the United States"; Marlowe, "Evidence of Selection for the Snorter Dwarf Gene in Cattle"; Preston, "Compact Cattle Genetics"; Preston and Willis, *Intensive Beef Production*.

41. Robert Hough, email to the author, August 28, 2020.

42. "The Indispensable Ingredient"; "Good, Don L."

43. Robert Hough, email to the author, August 28, 2020.

Epilogue

1. Robichaud, *Animal City*, 11–12.
2. "Fast Facts about Agriculture and Food"; Olmstead and Rhode, *Creating Abundance*, 262.
3. "Factsheet: USDA Coexistence Fact Sheets Corn"; "Factsheet: USDA Coexistence Fact Sheets Soybeans."
4. "The United States Meat Industry at a Glance."
5. Horowitz, "Making the Chicken of Tomorrow," 232.
6. See Purdy and Langemeier, "International Benchmarks for Corn Production" for the years 2018 and 2019.
7. Horowitz, *Putting Meat on the American Table*, 2.
8. "Saddle & Sirloin Club Portrait Collection Historical Overview."

Abott, Benjamin Vaughan. *A Digest of Reports of the United States Courts from the Beginning of the Year 1884 to December, 1888.* New York: Diossy, 1889.

"About Baby Beef." *Wallaces' Farmer*, January 13, 1905, 34.

Adams, Edward F. *The Modern Farmer in His Business Relations.* San Francisco: N. J. Stone, 1899.

"Adjusting Production to Consumption." *Wallaces' Farmer*, October 8, 1915, 6.

"Advancing Meat Prices Stir the Press." *American Meat Trade and Retail Butchers Journal* 15, no. 479 (1911): 6.

Alexander, A. S., and J. H. H. Alexander. "Farming: 'How to Make it Pay.'" *Better Farming* (1922): 5–6, 8.

Alexander, Jennifer Karns. *The Mantra of Efficiency: From Waterwheel to Social Control.* Baltimore, MD: Johns Hopkins University Press, 2008.

Allen, Lewis F. *American Cattle: Their History, Breeding, and Management.* New York: Orange Judd, 1879.

Almeroth-Williams, Thomas. *City of Beasts: How Animals Shaped Georgian London.* Manchester, UK: Manchester University Press, 2019.

"America to Feed World." *Duroc Bulletin and Livestock Farmer* 12, no. 305 (1917): 19.

Anderson, J. L. *Capitalist Pigs: Pigs, Pork, and Power in America.* Morgantown: West Virginia University Press, 2019.

——. *Industrializing the Corn Belt: Agriculture, Technology, and Environment, 1945–1972.* DeKalb: Northern Illinois University Press, 2009.

Anderson, Virginia DeJohn. *Creatures of Empire: How Domestic Animals Transformed Early America.* New York: Oxford University Press, 2004.

The Annual Register of a View of the History, Politics, and Literature, for the Year 1810. 2nd ed. London: Baldwin, Cradock, and Joy, 1825.

"Another Breeding Crate." *Breeder's Gazette* 80, no. 4 (1908): 180.

Armour, J. Ogden. *The Packers, the Private Car Lines, and the People.* Philadelphia: Henry Altemus, 1906.

Armour, Philip D. "The Relation of the Packing House to the Cattle Industry." In *Proceedings of the Third Annual Convention of the National Live Stock Association*, 208–12. Denver, CO: Smith-Brooks Printing, 1900.

Armsby, Henry P. "The Food Supply of the Future." In *Record of Proceedings of Annual Meeting, November, 1909*, 4–13. Champaign, IL: American Society of Animal Production, 1910.

Ashton, Eva. "Animal Husbandry: A Vocation for Women." *Shorthorn in America* 4, no. 10 (1919): 13–14.

"Average Weight of Livestock." In *"Our Year Book:" Telling Tables of the Livestock Trade for the Year 1901*, 19. Chicago: Chicago Daily Drovers Journal, 1902.

"Average Weight of Stock." In *"Our Year Book:" Telling Tables of the Livestock Trade and General Business for the Year 1912*, 9. Chicago: Chicago Daily Drovers Journal, 1913.

"The Babraham Southdown Flock." In *Live Stock Journal Almanac*, 332. London: Vinton, 1901.

"The Babraham Southdown Flock." In *Live Stock Journal Almanac*, 306. London: Vinton, 1910.

"Baby Beef." *Chicago Livestock World* 5, no. 298 (1904): 6.

"Baby Beef." *Chicago Livestock World* 9, no. 123 (1908): 2.

Bailey, Liberty Hyde. *The Country-Life Movement in the United States*. New York: Macmillan, 1913.

"Banks Form Syndicate to Supply Seed Corn." *Financier* 111, no. 8 (1918): 659.

"Barnyard Manure." *Prairie Farmer* 67, no. 13 (1895): 3.

Barrett, James R. *Work and Community in the Jungle: Chicago's Packinghouse Workers, 1894–1922*. Urbana: University of Illinois Press, 1987.

"Becoming a Breeder of Pure-Bred Stock." *Prairie Farmer* 73, no. 25 (1901): 7.

"Beef Cattle Breeding." *Prairie Farmer* 67, no. 15 (1895): 2.

"Beef Production in the Corn Belt: Feeding Cattle to Make Beef." *Wallaces' Farmer*, February 6, 1914, 5.

"Beef Production in the Corn Belt: Feeding Cattle to Sell Corn." *Wallaces' Farmer*, January 30, 1914, 5.

Bell, Alexander Graham. "A Few Thoughts Concerning Eugenics." In *Report of the Meeting Held at Washington D.C., January 28–30, 1908*, 208–14. Baltimore, MD: Kohn and Pollock, 1908.

Benson, R. R. "The Mental Processes of a Stock Judge." *Breeder's Gazette* 74, no. 3 (1918): 81.

Beresford, Rex. "Baby Beef Making Safest System for Corn Belt." *Prairie Farmer* 86, no. 2 (1914): 9, 39.

"Better Sires—Better Stock: That's the Slogan of a National Crusade of Great Importance." *Banker Farmer* 6, no. 10 (1919): 1–3.

"'Better Sires' Campaign Started by 'Uncle Sam.'" *Aberdeen-Angus Journal* 1, no. 7 (1919): 14.

"'Better Sires Demonstration Train' in Michigan." *Holstein-Friesian World* 18, no. 33 (1921): 26.

"The Biggest Problem on the American Farm." *Wallaces' Farmer*, November 15, 1907, 1318.

"A Big Reunion." *Swine World* 7, no. 8 (1919): 18.

Black, Edwin. *The War against the Weak: Eugenics and America's Campaign to Create a Master Race*. New York: Four Walls Eight Windows, 2003.

Black, W. H. "Beef and Dual-Purpose Cattle Breeding." In *Yearbook of the United States Department of Agriculture, 1936,* 863–86. Washington, DC: Government Printing Office, 1936.

Blanchette, Alex. *Porkopolis: American Animality, Standardized Life, and the Factory Farm.* Durham, NC: Duke University Press, 2020.

Blanke, David. *Sowing the American Dream: How Consumer Culture Took Root in the Rural Midwest.* Athens: Ohio University Press, 2000.

Bogue, Allan G. *From Prairie to Corn Belt: Farming on the Illinois and Iowa Prairies in the Nineteenth Century.* Chicago: University of Chicago Press, 1963.

Bowers, William L. *The Country Life Movement in America, 1900–1920.* Port Washington, NY: National University Publications, 1974.

Boyce, Lee. "Breeding Crate Saves Time and Trouble." *Berkshire World and Cornbelt Stockman* 13, no. 7 (1921): 6.

Boylan, Anthony Burke. "Amphitheatre Gets Its Final Curtain Call." *Chicago Tribune,* May 30, 1999.

"Breeders' Aid Is Sought." *Chicago Daily Tribune,* November 20, 1899.

"British Sheep Farming." *Wool Markets and Sheep* 8, no. 14 (1903): 6.

"British Sheep Farming." *Wool Markets and Sheep* 8, no. 15 (1903): 8.

Brown, Frank. "From the International Judge." *Shorthorn in America* 4, no. 4 (1919): 19.

Browning, H. E. "A Short History of Ohio Chief 8727A." *Swine World* 4, no. 12 (1917): 23.

Bryant, W. C. "Raising Bumper Grain Crops." *Prairie Farmer* 85, no. 5 (1912): 10.

"Bubbly Creek Dead, but Lives." *Chicago Daily Tribune,* May 20, 1915.

"Bubbly Creek's Doom Finally Decided Upon." *Chicago Daily Tribune,* October 17, 1919.

"Bubbly Creek Victim Lives." *Chicago Daily Tribune,* May 21, 1911.

"'Bubbly Creek's' Wonders Revealed to Investigators." *Chicago Daily Tribune,* October 15, 1905.

Burch, D. S. *Some Tested Methods for Livestock Improvement.* Washington, DC: Government Printing Office, 1925.

Burke, Tom, Kurt Schaff, and Jeremy Haag. *Central Illinois Aberdeen Angus Association Preview Show.* Smithville, MO: American Angus Hall of Fame, 2012.

"Campaign for Better Sires Now Includes Nearly 400,000 Head of Stock Enrolled." *Weekly News Letter of the U.S. Department of Agriculture* 8, no. 26 (1921): 10.

"The Carlots of Fat Cattle." *Breeder's Gazette* 80, no. 23 (1921): 854–55.

Carney, Judith. *Black Rice: The African Origins of Rice Cultivation in the Americas.* Cambridge, MA: Harvard University Press, 2001.

"Cattle." In *"Our Year Book:" Telling Tables of the Livestock Trade for the Year 1901,* 6. Chicago: Chicago Daily Drovers Journal, 1902.

"Cattle Feeding on Increase." *Wallaces' Farmer,* February 1, 1936, 10.

Chafe, William Henry. *The American Woman: Her Changing Social, Economic, and Political Roles, 1920–1970.* New York: Oxford University Press, 1972.

Chandler, F. M. "Breeding for a Purpose." *Shepherd's Criterion* 15, no. 12 (1905): 5.

"Changes on the Range." *National Provisioner* 26, no. 17 (1902): 13.

"Charles Robert Darwin." *American Breeders Magazine* 1, no. 1 (1910): 9–10.

"Chicago's Food the Best: Secretary of Agriculture Wilson Gives Opinion." *Swift & Company Yearbook.* Chicago: Union Stock Yards, 1908.

"Chicago Selected." *Wool Markets and Sheep* 11, no. 14 (1901): 17.

"Chicago Tribune." *National Provisioner* 26, no. 17 (1902): 15.

Christy, Ralph D., and Lionel Williamson, eds. *A Century of Service: Land-Grant Colleges and Universities, 1890–1990.* New Brunswick, NJ: Transaction, 1992.

Clarke, W. J. *Fitting Sheep for the Show Ring and Market.* Chicago: Draper, 1900.

Clemen, Rudolf Alexander. *The American Livestock and Meat Industry.* New York: Ronald Press, 1923.

Coburn, J. D. *Swine Husbandry.* New York: Orange Judd, 1919.

Coffey, W. C. "Market Classes and Grades of Sheep." *University of Illinois Agricultural Experiment Station* 129 (1908): 577–635.

Cohen, Benjamin R. *Notes from the Ground: Science, Soil, and Society in the American Countryside.* New Haven, CT: Yale University Press, 2009.

"The Coming International." *Shepherd's Criterion* 15, no. 11 (1905): 7.

"Concrete for the Farm." *Agricultural Digest* 2, no. 6 (1917): 713, 725.

Conkin, Paul K. *A Revolution Down on the Farm: The Transformation of American Agriculture since 1929.* Lexington: University Press of Kentucky, 2008.

Connor, L. G. "A Brief History of the Sheep Industry in the United States." *American Sheep Breeder and Wool Growers* 42, no. 9 (1922): 462–66.

Cook, O. F. "City and Country: Effects of the Human Environment on the Progress of Human Civilization," pt. 1. *Journal of Heredity* 12, no. 3 (1921): 112–16.

——. "City and Country: Effects of the Human Environment on the Progress of Human Civilization," pt. 2. *Journal of Heredity* 12, no. 4 (1921): 167–73.

Cooper, Richard Powell. "The Export Trade in Pedigree Stock." In *Live Stock Journal Almanac* 108–10. London: Vinton, 1910.

Cott, Nancy F. *The Grounding of Modern Feminism.* New Haven, CT: Yale University Press, 1988.

Coulter, J. L. "Agricultural Development in the United States, 1900–1920." *Quarterly Journal of Economics* 27, no. 1 (1912): 1–26.

"The Court Decision Governing Importations of Breeding Stock." *Breeder's Gazette* 11, no. 21 (1887): 939.

Craft, W. C. "Fifty Years of Progress in Swine Breeding." *Journal of Animal Science* 17, no. 4 (1958): 960–80.

Craig, John A. *Judging Live Stock.* Des Moines, IA: Kenyon Printing and Manufacturing, 1901.

Craig, R. A. *A Brief Practical Treatise on Veterinary Medicine: Diseases of Farm Animals.* Chicago: Rand, McNally, 1902.

Crissey, Forrest. "First Aids to Farmers." *Country Gentleman* 84, no. 15 (1919): 6–7, 40, 42.

Cronon, William. *Nature's Metropolis: Chicago and the Great West.* New York: Norton, 1991.

Crosby, Alfred. *Ecological Imperialism: The Biological Expansion of Europe, 900–1900.* New York: Cambridge University Press, 1986.

——. *The Columbian Exchange: Biological and Cultural Consequences of 1492.* Westport, CT: Greenwood, 1972.

Curtis, Robert S. *The Fundamentals of Live Stock Judging and Selection.* 2nd ed. Philadelphia and New York: Lea and Febiger, 1920.

Curtiss, Charles F. "An American on British Stock." In *Live Stock Journal Almanac,* 100–102. London: Vinton, 1900.

"Dairy Demonstration Tour in Michigan." *Holstein-Friesian World* 18, no. 29 (1921): 3022.

Danbom, David B. *The Resisted Revolution: Urban American and the Industrialization of Agriculture, 1900–1930.* Ames: Iowa State University Press, 1979.

Daniel, Pete. *Breaking the Land: The Transformation of Cotton, Tobacco, and Rice Cultures since 1880.* Urbana: University of Illinois Press, 1986.

Darlow, A. E. "Fifty Years of Livestock Judging." *Journal of Animal Science* 17, no. 4 (1958): 1058–63.

Davenport, C. S. *Heredity in Relation to Eugenics.* New York: Henry Holt, 1911.

Davenport, Eugene. "Scientific Farming." *Annals of the American Academy of Political and Social Science* 40, no. 1 (1912): 45–50.

"Death of Prof. John A. Craig." *Breeder's Gazette* 58, no. 7 (1910): 270.

"Deaths." *University Journal* 17, no. 2 (1921): 32.

"Decline in Rural Population." *Wallaces' Farmer*, October 13, 1905, 1194.

"Decline in Soil Fertility." *Chicago Livestock World* 14, no. 45 (1913): 2.

"Dedication of the New Building." *Breeder's Gazette* (1903): 1114.

De Loach, R. J. H. *Armour's Handbook of Agriculture.* Chicago: Armour, 1921.

——. "Beef Cattle." *Armour's Bureau of Agricultural Research and Economics* 5 (1918).

——. "The Tale of Two Steers." *Farm Boys' and Girls' Leader and Club Achievements* 2, no. 8 (1920): 3.

De Loach, R. J. H., and H. A. Phillips. *Progressive Sheep Raising.* Chicago: Armour's Bureau of Agricultural Research and Economics, 1918.

"Demand for Baby Beef." *Chicago Livestock World* 8, no. 276 (1907): 4.

Dietrich, William. "Market Classes and Grades of Swine." *University of Illinois Agricultural Experiment Station* 97 (1904): 419–63.

Donahue, Brian. *The Great Meadow: Farmers and the Land in Colonial Concord.* New Haven, CT: Yale University Press, 2004.

Eleventh Annual Report of the Chicago Junction Railways and Union Stock Yards Company. Chicago: Chicago Junction Railways and Union Stock Yards Company, 1902.

"Enterotoxaemia Can Hit Healthy Lambs." *Farmers' Weekly Review* 36, no. 41 (1957): 2.

Evans, Robert J. *History of the Duroc.* Chicago: James J. Doty, 1918.

Evvard, John M. "Producing Baby Beeves on Corn Belt Farms." *Prairie Farmer* 90, no. 26 (1918): 9, 32.

"Exhausting the Soil." *Chicago Livestock World* 6, no. 54 (1910): 2.

"Experts Tell Radio World Romantic Story of Live Stock." *Meat and Live Stock Digest* 4, no. 7 (1924): 2, 4.

"Facts about Soil Fertility." *Prairie Farmer* 85, no. 5 (1912): 12.

"Factsheet: USDA Coexistence Fact Sheets Corn." United States Department of Agriculture, https://www.usda.gov/sites/default/files/documents/coexistence-corn -factsheet.pdf.

"Factsheet: USDA Coexistence Fact Sheets Soybeans." *United States Department of Agriculture,* https://www.usda.gov/sites/default/files/documents/coexistence -soybeans-factsheet.pdf.

Faragher, John Mack. *Sugar Creek: Life on the Illinois Prairie.* New Haven, CT: Yale University Press, 1986.

"Fast Facts about Agriculture and Food." American Farm Bureau Federation, https://
www.fb.org/newsroom/fast-facts#:~:text=One%20U.S.%20farm%20feeds%20
166,than%20what%20is%20now%20produced.

Fenton, Sarah. "International Amphitheater." *Encyclopedia of Chicago*, http://www
.encyclopedia.chicagohistory.org/pages/647.html.

Ferleger, Louis. "Arming American Agriculture for the Twentieth Century: How the
USDA's Top Managers Promoted Agricultural Development." *Agricultural History*
74, no. 2 (2000): 211–26.

Fifth Annual Report of the Chicago Junction Railways and Union Stock Yards Company.
Chicago: Chicago Junction Railways and Union Stock Yards Company, 1895.

"Finds Scrub Bull Guilty." *Meat and Livestock Digest* 9, no. 12 (1928): 3.

Finlay, Mark. *Growing American Rubber: Strategic Plants and the Politics of National
Security.* New Brunswick, NJ: Rutgers University Press, 2009.

Fitzgerald, Deborah. *The Business of Breeding: Hybrid Corn in Illinois, 1890–1940.*
Ithaca, NY: Cornell University Press, 1990.

———. *Every Farm a Factory: The Industrial Ideal in American Agriculture.* New Haven,
CT: Yale University Press, 2003.

Flock Book of the National Cheviot Sheep Society. Springfield: Illinois State Register,
1898.

Flock Register of the American Cheviot Sheep Society. Fayetteville, NY: American
Cheviot Sheep Breeders' Association, 1901.

"Food Production and Conservation Committee." *Report on the Work of the State
Council of Defense of Illinois* 1, no. 5 (1918): 10–11.

"Food Program for Illinois." *Swine World* 5, no. 8 (1918): 6, 13.

"For Chicago Live Stock Show: First Steps to Plan an International Exhibition to Be
Taken at the Yards Today." *Chicago Daily Tribune*, November 24, 1899.

"The Foreign Breed of Sheep." In *Sheep: Their Breeds, Management, and Diseases*,
108–36. London: Baldwin and Craddock, 1837.

"The Fourth International Grain and Hay Show." In *Nineteenth Annual Report of the
Indiana Corn Growers' Association*, 58–65. Indianapolis, IN: William. B. Burford,
1919.

"Fresh Meats: The Export Trade in Dressed Beef, Pork, Mutton, Etc.—Why Live
Animals Should No Longer Be Transported Long Distances." *Chicago Daily Tribune*,
May 6, 1877.

Gardner, Bruce L. *American Agriculture in the Twentieth Century: How it Flourished
and What It Cost.* Cambridge, MA: Harvard University Press, 2002.

Gay, Carl Warren. *The Principles and Practice of Judging Live-Stock.* New York:
Macmillan, 1914.

"General Review 1901." In *"Our Year Book": Telling Tables of the Livestock Trade for the
Year 1901*, 5–9. Chicago: Chicago Daily Drovers Journal, 1902.

"Glad Mourners for Bubbly Creek." *Chicago Daily Tribune*, May 21, 1915.

Glenna, Leland L., Margaret A. Gollnick, and Stephen S. Jones. "Eugenic Opportunity
Structures: Teaching Genetic Engineering at U.S. Land-Grant Universities since
1911." *Social Studies of Science* 37, no. 2 (2007): 281–96.

Gobble, C. F. "Indiana and Purdue Win Again." *Purdue Agriculturalist* 14, no. 4 (1921):
139–40.

Golloway, William. "Soil Fertility." *Farm Home* 40, no. 382 (1914): 9, 17.

Good, Don L." North American Livestock Exposition, https://naile.s3.amazonaws.com /2021/06/6-Portraits-F-I.pdf.

"Good Enough for the Breeder and Cheap Enough for the Farmer." *Berkshire World and Cornbelt Stockman* 4, no. 9 (1912): 28.

"Good Management Protects Calves from Coccidiosis." *Farmers' Weekly Review* 20, no. 29 (1942).

"The Gospel of Improvement." *Chicago Livestock World* 12, no. 108 (1911): 2.

Grand, W. Joseph. *Illustrated History of the Union Stock Yards: A Sketch-Book of Familiar Faces and Places at the Yards.* Chicago: Thomas Knapp, 1896.

"The Great Livestock Exposition." *Railway Journal* 23, no. 12 (1917): 8.

"Gregor Johann Mendel." *American Breeders Magazine* 1, no. 1 (1910): 10–13.

Goldman, Michael. *Imperial Nature: The World Bank and Struggles for Social Justice in the Age of Globalization.* New Haven, CT: Yale University Press, 2006.

"A Good Record." *Wool Markets and Sheep* 14, no. 1 (1904): 29.

Goodwin, John S. "The Performance Record of Angus Cattle: A Restatement of the Case for the Breed, With a Digest of Evidence." *Breeder's Gazette* 70, no. 25 (1916): 1206–7.

Hackleman, J. C. *History: International Crop Improvement Association, 1919–1961.* Clemson, S.C.: International Crop Improvement Association, 1961.

Hall, Louis D. "Market Classes and Grades of Meat." *University of Illinois Agricultural Experiment Station* 147 (1910): 147–290.

Hammond, J. W. "Wool Investigations at the Ohio Experiment Station." In *Record of Proceedings of the Annual Meeting, November, 1914,* 37–44. Champaign, IL: American Society of Animal Production, 1915.

Harding, T. Swann. *Two Blades of Grass: A History of Scientific Developments in the U.S. Department of Agriculture.* Norman: University of Oklahoma Press, 1947.

Hartman, F. B. "The Cheviot." *Wool Markets and Sheep* 12, no. 10 (1902): 6.

"Hartman's Cheviots." *Wool Markets and Sheep* 11, no. 7 (1901): 2.

"Has the International Made Good?" *Agricultural Digest* 3, no. 3 (1921): 37–38.

Hayes, O. M. "Baby Beef Production." *Breeder's Gazette* 63, no. 4 (1913): 194.

Heath, A. S. "For More and Better Meat: How to Increase and Improve Our Meat Animals." *National Provisioner* 39, no. 1 (1908): 17.

———. "For More and Better Meat: How to Increase and Improve Our Meat Animals." *National Provisioner* 39, no. 6 (1908): 17.

——— "For More and Better Meat: How to Increase and Improve Our Meat Animals." *National Provisioner* 39, no. 9 (1908): 17.

———. "For More and Better Meat: Proper Cross Breeding Suggested as Cure for Disease." *National Provisioner* 39, no. 5 (1908): 17.

Helmer, Richard Bryan. *James and Alvin Sanders: Livestock Journalists of the Midwest.* Bryn Mawr, PA: Dorrace, 1985.

Henry, W. A., and F. B. Morrison. *Feeds and Feeding Abridged: The Essentials of the Feeding, Care, and Management of Farm Animals, Including Poultry.* Madison, WI: Henry-Morrison, 1921.

———. *Feeds and Feeding: A Handbook for the Student and Stockman.* Madison, WI: Henry-Morrison, 1920.

"High Prices." *American Meat Trade and Retail Butchers Journal* 15, no. 479 (1911): 5.

"High Prices for Pedigree Stock." *Prairie Farmer* 70, no. 46 (1898): 8.

Hill, Howard Copeland. "The Development of Chicago as a Center of the Meat Packing Industry." *Mississippi Valley Historical Review* 10, no. 3 (1923): 253–73.

Hinman, Robert B., and Robert B. Harris. *The Story of Meat*. Chicago: Swift, 1939.

Hobsbawm, Eric. *The Age of Extremes: A History of the World, 1914–1991*. New York: Vintage, 1994.

Hofstadter, Richard. *The Age of Reform: From Bryan to F.D.R.* New York: Knopf, 1955.

"Hopkins Addresses Rural Teachers." *Prairie Farmer* 78, no. 37 (1906): 2–3.

Hopkins, Cyril G. "Plant Food in Relation to Soil Fertility." *University of Illinois Agricultural Experiment Station Circular* 155 (1912): 1–10.

——. *Soil Fertility and Permanent Agriculture*. Boston: Ginn, 1910.

——. "The Story of the Soil." *Prairie Farmer* 83, no. 11 (1911): 15, 19.

Horowitz, Roger. "Making the Chicken of Tomorrow." In *Industrializing Organisms: Introducing Evolutionary History*, ed. Susan R. Schrepfer and Philip Scranton, 215–35. New York: Routledge, 2004.

——. *Putting Meat on the American Table: Taste, Technology, Transformation*. Baltimore, MD: Johns Hopkins University Press, 2006.

"How to Increase Pork Supply." *Berkshire World and Cornbelt Stockman* 10, no. 1 (1918): 43.

"How to Keep Boys on the Farm." *Wallaces' Farmer*, June 26, 1908, 2.

Hurt, R. Douglas. *Problems of Plenty: The America Farmer in the Twentieth Century*. Chicago: Ivan R. Dee, 2002.

"The Ideal Corn Belt Farm." *Wallaces' Farmer*, March 27, 1914, 4.

Igler, David. *Industrial Cowboys: Miller & Lux and the Transformation of the Far West, 1850–1920*. Berkeley: University of California Press, 2001.

"Indiana Leads in Winnings." *Purdue Agriculturalist* 17, no. 4 (1923): 63, 72.

"The Indispensable Ingredient: Angus Blood Dominates Chicago Steer Show." *Angus Journal* 51, no. 6 (1970): 44–45, 128.

"The International Answers the Nation's Call." *American Breeder* 11, no. 5 (1917): 16.

"International Exposition." In *"Our Year Book:" Telling Tables of the Livestock Trade for the Year 1901*, 9. Chicago: Chicago Daily Drovers Journal, 1902.

"International Live-Stock Exposition." *Breeder's Gazette* (1900): 709.

"International Livestock Exposition." *Ohio Farmer* 144, no. 24 (1919): 3, 24, 30, 33.

"International Live Stock Exposition." *Wool Markets and Sheep* 11, no. 24 (1901): 18.

"International Live Stock Exposition: The Exponent of a Great Movement for Improvement of the Domestic Animals of the United States." *Opportunities of To-Day* 3, no. 6 (1907): 29–38.

"International Live Stock Show." *Duroc Bulletin and Livestock Farmer* 12, no. 303 (1917): 42.

"The International Livestock Show." *Prairie Farmer* 91, no. 27 (1919): 12.

"The International Swine Show." *Swine World* 1, no. 10 (1913): 3–5, 8.

Irwin, Lyndon. "Agricultural Events at the 1904 St. Louis World's Fair." http://www.lyndonirwin.com/1904fair.htm.

"'Jim' Poole Says the International Has Missed Its Calling: Has It?" *Aberdeen-Angus Journal* 2, no. 12 (1921): 9, 24–25.

Jordan, David Starr. "Report of the Committee on Eugenics." In *Report of the Meeting Held at Washington D.C., January 28–30, 1908*, 201–8. Baltimore, MD: Kohn and Pollock, 1908.

Jordan, John. *Machine-Age Ideology: Social Engineering and American Liberalism, 1911–1939.* Chapel Hill: University of North Carolina Press, 1994.

Jordan, Terry. *North American Cattle-Ranching Frontiers: Origins, Diffusion, and Differentiation.* Albuquerque: University of New Mexico Press, 1993.

J. W. C. "The Value of Pedigree." *Prairie Farmer* 72, no. 46 (1900): 3.

Kennedy, David M. *Over Here: The First World War and American Society.* New York: Oxford University Press, 1980.

Kerr, Norwood Allen. *The Legacy: A Centennial History of the State Agricultural Experiment Stations, 1887–1987.* Columbia: Missouri State Experiment Station, 1987.

Kimmelman, Barbara. "The American Breeders' Association: Genetics and Eugenics in an Agricultural Context, 1903–1913." *Social Studies of Science* 13, no. 2 (1983): 163–204.

King, F. H. *The Soil: Its Nature, Relations, and Fundamental Principles of Management.* New York: Macmillan, 1895.

Kipling, Rudyard. *American Notes.* New York: Frank F. Lovell, 1899.

Kline, Ronald. *Consumers in the Country: Technology and Social Change in Rural America.* Baltimore, MD: Johns Hopkins University Press, 2000.

Knapp, Joseph G. "A Review of Chicago Stock Yards History." *University Journal of Business* 2, no. 3 (1924): 331–46.

Kraemer, H. "Effects of Inbreeding." *Journal of Heredity* 5, no. 5 (1914): 226–34.

Krueck, W. B. "Topping the Market with Baby Beef." *Breeder's Gazette* 76, no. 16 (1919): 778, 780.

Kujovich, Mary Yeager. "The Refrigerator Car and the Growth of the American Dressed Beef Industry." *Business History Review* 44, no. 4 (1970): 460–82.

Lambert, George William. *A Trip through the Union Stock Yards and Slaughter Houses.* Chicago: Hamblin, [ca. 1900].

Langley, Lester D. *The Banana Wars: United States Intervention in the Caribbean, 1898–1934.* Lexington: University Press of Kentucky, 1983.

"Largest Exposition Building in World Nearing Completion in Chicago." *Chicago Daily Tribune,* November 5, 1905.

Lawrence, Lee E. "The Wisconsin Ice Trade." *Wisconsin Magazine of History* 48, no. 4 (1965): 257–67.

Lectures on the Results of the Exhibition Delivered before the Society of Arts, Manufactures, and Commerce. London: David Bogue, 1852.

Leech, Harper, and John Charles Carroll. *Armour and His Times.* New York: Appleton-Century, 1938.

Levario, Miquel Antonio. *Militarizing the Border: When Mexicans Became the Enemy.* 2nd ed. College Station: Texas A&M University Press, 2015.

Levins, Richard. *William Cochrane and the American Family Farm.* Lincoln: University of Nebraska Press, 2000.

"Line Up in War on Bubbly Creek." *Chicago Daily Tribune,* May 10, 1915.

"Live Stock Exposition: The Union Stock Yards Plan for an International Event." *Washington Post,* November 20, 1899.

"Livestock Breeding at the Crossroads." In *Yearbook of the United States Department of Agriculture, 1936*, 831–62. Washington, DC: Government Printing Office, 1936.

"Livestock Farming Conserves Soil." *Prairie Farmer* 91, no. 29 (1919): 11, 33.

Live Stock Journal Almanac. London: Vinton, 1902.

"Live Stock Show Aid to Industry." *Chicago Daily Tribune*, December 2, 1908.

Lloyd-Jones, Orren. "What Is a Breed?" *Journal of Heredity* 6, no. 12 (1915): 531–37.

Lyon, T. Lyttleton, and Elmer O. Fippin. *The Principles of Soil Management*. New York: Macmillan, 1912.

Magdoff, Fred, and Brian Tokar, eds. *Agriculture and Food in Crisis: Conflict, Resistance, and Renewal*. New York: Monthly Review Press, 2010.

"Maintenance of Fertility." *Wallaces' Farmer*, March 5, 1909, 3.

"Maple Grove Flock Doing Well." *Wool Markets and Sheep* 14, no. 6 (1904): 11.

Marcus, Alan I., ed. *Science as Service: Establishing and Reformulating American Land-Grant Universities, 1865–1930*. Tuscaloosa: University of Alabama Press, 2015.

——, ed. *Service as Mandate: How American Land-Grant Universities Shaped the Modern World, 1920–2015*. Tuscaloosa: University of Alabama Press, 2015.

Marlowe, Thomas J. "Evidence of Selection for the Snorter Dwarf Gene in Cattle." *Journal of Animal Science* 23, no. 2 (1964): 454–60.

Marshall, Duncan. "A Study in Scotch Pedigree." *Shorthorn World and Farm Magazine* 3, no. 22 (1919): 4–8, 68–71.

Marshall, F. R. *Breeding Farm Animals*. Chicago: Breeder's Gazette, 1912.

Martin, Oscar Baker. *The Demonstration Work: Dr. Seaman A. Knapp's Contribution to Civilization*. Boston: Stratford, 1921.

Marx, Leo. *The Machine in the Garden: Technology and the Pastoral Ideal in America*. New York: Oxford University Press, 1964.

Maxwell, Cyrus H. "An Analysis of the Results of the Steer Carcass Contest at the International Livestock Exposition, 1908–1923." *Journal of Animal Science* (October 1924): 132–35.

McCartney, H. E. "Merry Monarch a Great Steer." *Field Illustrated* 28, no. 1 (1918): 19, 60, 62.

McCelland, Peter D. *Sowing Modernity: America's First Agricultural Revolution*. Ithaca, NY: Cornell University Press, 1997.

McFarlane, Thomas. *The American Aberdeen-Angus Herd-Book*. Davenport, IA: Egbert, Fidlar, and Chambers, 1901.

McGavock, W. C. "The Trend of the Times in the American Hereford Trade." *Breeder's Gazette* 70, no. 25 (1916): 1190–91.

McGee, Harold. *On Food and Cooking: The Science and Lore of the Kitchen*. New York: Scribner, 2004.

McGregor, Alexander Campbell. *Counting Sheep: From Open Range to Agribusiness on the Columbia Plateau*. Seattle: University of Washington Press, 1982.

McMichael, Philip, ed. *Food and Agrarian Orders in the World-Economy*. Westport, CT: Praeger, 1995.

McWorter, V. O. "Equipment for Farm Sheep Raising." *United States Department of Agriculture Farmers' Bulletin*. 810 (1917): 3–27.

"The Meat Situation." *National Provisioner* 26, no. 16 (1902): 11–12.

Merchant, Carolyn. *The Death of Nature: Women, Ecology, and Scientific Revolution.* San Francisco: Harper and Row, 1980.

Merrill, Karen R. *Public Lands and Political Meaning: Ranchers, the Government, and the Property between Them.* Berkeley: University of California Press, 2002.

"Michigan for Better Dairy Stock." *Co-operative Manager and Farmer* 11, no. 4 (1921): 50.

Miles, Manly. *Stock-Breeding: A Practical Treatise on the Application of the Laws of Development and Heredity to the Improvement and Breeding of Domestic Animals.* New York: Appleton, 1889.

Miller, Donald L. *City of the Century: The Epic of Chicago and the Making of America.* New York: Simon and Schuster, 1996.

Minster, Leo. "Hereford Claims from England." *Breeder's Gazette* 76, no. 8 (1919): 309.

Miraldi, Robert. *The Pen Is Mightier: The Muckraking Life of Charles Edward Russell.* New York: Palgrave Macmillan, 2003.

"Mr. F. B. Hartman." *Wool Markets and Sheep* 12, no. 16 (1902): 17.

Mohler, John R. *"Better Sires—Better Stock": Plan of Nation-Wide Crusade to Improve Quality of Live Stock through Use of Good Pure-Bred Sires.* Washington, DC: Government Printing Office, 1919.

Mumford, Herbert W. *Beef Production.* Urbana, Illinois: published by the author, 1907.

——. "Market Classes and Grades of Cattle with Suggestions for Interpreting Market Quotations." *University of Illinois Agricultural Experiment Station* 78, (1902): 367–433.

——. "Mumford on Beef Production." *Chicago Livestock World* 7, no. 270 (1906): 6.

Myers, John F. "Handling the Sows during the Breeding Season." *Berkshire World and Cornbelt Stockman* 6, no. 11 (1914): 6.

Nash, Richard. "Cattle Paper and the Changing Conditions of the Live Stock Trade." In *Proceedings of the Fourth Annual Convention of National Live Stock Association,* 388–92. Denver, CO: Smith-Brooks, 1901.

"A Nation of Soil Robbers." *Wallaces' Farmer,* August 30, 1907, 938.

"New International Live Stock Exposition Building: The Largest of its Kind in the World." *Shepherd's Criterion* 15, no. 12 (1905): 30.

"A New Name for the Scrubs." *Weekly News Letter of the U.S. Department of Agriculture* 7, no. 41 (1920): 2.

"The 1917 'International.'" *Agricultural Digest* 2, no. 6 (1917): 727.

Ninth Annual Report of the Chicago Junction Railways and Union Stock Yards Company. Chicago: Chicago Junction Railways and Union Stock Yards Company, 1900: 1–15

Obrecht, Rufus C. "Market Classes and Grades of Horses and Mules." *University of Illinois Agricultural Experiment Station* 122 (1908): 93–186.

O'Brien, John. *Through the Chicago Stock Yards; A Handy Guide to the Great Packing Industry.* Chicago: Rand, McNally, 1907.

Official Catalogue: International Live Stock Exposition. Chicago: Union Stock Yards, 1910.

Olmstead, Alan L., and Paul W. Rhode. *Creating Abundance: Biological Innovation and American Agricultural Development.* New York: Cambridge University Press, 2008.

O'Neill, William L. *Everyone Was Brave: The Rise and Fall of Feminism in America.* Chicago: Quadrangle, 1969.

"Orders Packers to Dig: Drainage Board Wants 'Bubbly Creek' Dredged." *Chicago Daily Tribune*, May 30, 1906.

"Our Beef and Its Maligners." *National Provisioner* 26, no. 14 (1902): 19.

"Our Beef Supply." In *Yearbook of the United States Department of Agriculture, 1921*, 227–322. Washington, DC: Government Printing Office, 1922.

Outline for Conducting a Scrub Sire Trial. Washington, DC: Government Printing Office, 1924.

"Overeating Disease Is Serious Threat to Feeder Lambs." *Farmers' Weekly Review* 26, no. 45 (1947): 1.

Paarlberg, Robert. *Food Politics: What Everyone Needs to Know*. New York: Oxford University Press, 2010.

Pacyga, Dominic A. *Slaughterhouse: Chicago's Union Stockyard and the World It Made*. Chicago: University of Chicago Press, 2015.

Pate, J'Nell L. *America's Historic Stockyards: Livestock Hotels*. Fort Worth: Texas Christian University Press, 2005.

"Paved Feedlots Found Profitable." *Farmers' Weekly Review* 35, no. 17 (1956): 4.

Pawley, Emily. *The Nature of the Future: Agriculture, Science, and Capitalism in the Antebellum North*. Chicago: University of Chicago Press, 2020.

"Pedigrees of Live Stock." *Wallaces' Farmer* 26, July 5, 1901, 836.

"Permanent Buildings." *Prairie Farmer* 88, no. 23 (1916): 25.

Phillips, Sarah T. *This Land, This Nation: Conservation, Rural America, and the New Deal*. New York: Cambridge University Press, 2007.

Pickett, John E. "White Faces: They are Thickest Round Kansas City, Which Has Become the Herefordshire of America." *Country Gentleman* 84, no. 3 (1919): 5.

Piggott, J. R. *Palace of the People: The Crystal Palace at Sydenham, 1854–1936*. Madison: University of Wisconsin Press, 2004.

Plans of Farm Buildings for Western States. Washington, DC: United States Department of Agriculture, 1939.

Plumb, Charles S. *Beginnings in Animal Husbandry*. St. Paul, MN: Webb, 1921.

———. "The Function of the Constructive Breeder of Registered Live Stock." *American Breeder* 9, no. 7 (1915): 11.

———. "Judging Stock at the Colleges." *Farmer's Guide* (December 1901).

———. "Students' Judging Contest." *Breeder's Gazette* 45, no. 4 (1904): 152–53.

———. "Students' Judging Contests Again." *Breeder's Gazette* 42, no. 6 (1902): 208.

———. *A Study of Farm Animals*. St. Paul, MN: Webb, 1922.

———. "Teaching Animal Selection." In *Report of the Meeting Held at Columbus, Ohio, January 15–18*, 83–95. Washington, DC: American Breeders' Association, 1907.

———. *Types and Breeds of Farm Animals*. Boston: Ginn, 1906.

———. *Types and Breeds of Farm Animals*. Revised ed. Boston: Athenaeum, 1920.

———. "A Type of Breed: A Plea for Higher Attainments in the Breeding of Stock." *Farmers' Magazine* 1, no. 1 (1894).

"Points on Treatment of Herd Boars." *American Swineherd* 39, no. 4 (1922): 11–12.

Poole, James E. "Beef Production on New Basis." *Producer* 3, no. 9 (1922): 24–25.

———. "Craze for Light Cattle." *Producer* 3, no. 7 (1921): 21.

———. "The International Anniversary Show: A Review of the Origin and Development and an Appreciation of the Influence of the International Live Stock Exposition,

Whose Twentieth Birthday Will Be Fittingly Celebrated in Chicago Next Week."
     Breeder's Gazette 76, no.22 (1919): 1147–48.

——. "The Twentieth International: Retrospective View of the Needs and Conditions
     that Brought Into Being the World's Most Conspicuous Live Stock Show." *Shorthorn
     World* 4, no. 18 (1919): 13–14.

——. "Why Beef Consumption Is Lagging." *Producer* 3, no. 12 (1922): 13.

——. "Young Cattle the Most Profitable." *Breeder's Gazette* 80, no. 12 (1921): 407.

"Praises Show of Livestock: Secretary Wilson Says It Is a Magnificent Exposition." *San
     Francisco Call*, December 2, 1902.

Prasad, Monica. *The Land of Too Much: American Abundance and the Paradox of
     Poverty*. Cambridge, MA: Harvard University Press, 2012.

*Premium List for the Eleventh Annual American Fat Stock Show, and the American Live
     Stock Show to be Held in the Exposition Building, Chicago, November 13–24, 1888,
     under the Auspices of the Illinois State Board of Agriculture*. Chicago: J. M. W. Jones,
     1888.

"President Wilson Enrolls Flock in Better Sires Campaign." *Weekly News Letter of the
     U.S. Department of Agriculture* 7, no. 41 (1920): 1.

Preston, R. L. "Compact Cattle Genetics." In *Stetson, Pipe and Boots: Colorado's
     Cattleman Governor*, 119–30. Victoria, BC: Trafford, 2006.

Preston, Thomas R., and Malcolm. B. Willis. *Intensive Beef Production*. Oxford, UK:
     Pergamon Press, 1970.

"Prize Stock Given Finishing Touch: All Now Ready for Formal Opening of Interna-
     tional Exposition Today." *Chicago Daily Tribune*, November 29, 1909.

"The Production of Baby Beef." *Chicago Livestock World* 16, no. 73 (1915): 3.

"A Productive Stock Farm." *Prairie Farmer* 90, no. 29 (1918): 9.

"Prof. John A. Craig." *Breeder's Gazette* 58, no. 8 (1910): 299.

Purdy, Rachel, and Michael Langemeier. "International Benchmarks for Corn Produc-
     tion." *Farmdoc Daily* 8, no. 100 (2018), https://farmdocdaily.illinois.edu/2018/06
     /international-benchmarks-for-corn-production-3.html.

——. "International Benchmarks for Soybean Production." *Farmdoc Daily* 9, no. 94
     (2019), http://www.farmdocdaily.illinois.edu/2019/05/international-benchmarks
     -for-soybean-production-3.html.

"Pure Bred Sheep." *Wool Markets and Sheep* 5, no. 10 (1900): 16.

"Question of Pedigree." *Wallaces' Farmer*, June 30, 1905, 831.

"Race Genetics Problems." *American Breeders Magazine* 2, no. 3 (1911): 230–32.

Rasmussen, Wayne D. ed. *Readings in the History of American Agriculture*. Urbana:
     University of Illinois Press, 1960.

——. *Taking the University to the People: Seventy-Five Years of Cooperative Extension*.
     Ames: Iowa State University, 1989.

Ray, S. H. *The Production of Baby Beef*. Washington, DC: United States Department of
     Agriculture, 1917.

"The Real Meat Facts." *National Provisioner* 26, no. 14 (1902): 22–23.

"Registry and Transfer Fees." *Wallaces' Farmer*, March 2, 1917, 9.

"Remarkable Classes of Beef Breeding Cattle." *Breeder's Gazette* 82, no. 24 (1922):
     838–40.

Report of the Commission on Country Life. New York: Sturgis and Walton, 1917.

"Report of the Committee on Terminology." In *Record of Proceedings of Annual Meeting, November, 1914*, 115–23. Champaign, IL: American Society of Animal Production, 1915.

Review of the First International Live Stock Exposition. Chicago: Union Stock Yard and Transit Company, 1900.

A Review of the International Live Stock Exposition: A Great Movement for Improvement of the Domestic Animals of the United States. Chicago: Union Stock Yard and Transit Company, 1913, 1916, 1917, 1918, 1919, 1921, 1922, 1948.

Richardson, Heather Cox. *How the South Won the Civil War.* New York: Oxford University Press, 2020.

Rifkin, Jeremy. *Beyond Beef: The Rise and Fall of the Cattle Culture.* New York: Dutton, 1992.

Ritchie, Harlan. "From Big to Small to Big to Small: Our History of Cattle Breeding from 1742 to Today," Pt. 1. *On Pasture*, July 4, 2016. https://onpasture.com/2016/07/04/from-big-to-small-to-big-to-small-a-pictorial-history-of-how-weve-changed-what-cattle-look-like.

———. "From Big to Small to Big to Small: A Pictorial History of Cattle Changes Over the Years," Pt. 2. *On Pasture*, July 11, 2016. https://onpasture.com/2016/07/11/from-big-to-small-to-big-to-small-part-2-of-a-pictorial-history-of-cattle-changes-over-the-years.

———. "From Big to Small to Big to Small: A Pictorial History of Cattle Over the Years," Pt. 3. *On Pasture*, July 18, 2016. https://onpasture.com/2016/07/18/from-big-to-small-to-big-to-small-part-3-of-a-pictorial-history-of-cattle-over-the-years.

Ritvo, Harriet. *The Animal Estate: The English and Other Creatures in the Victorian Age.* Cambridge, MA: Harvard University Press, 1987.

"Robert Bakewell." *American Breeders Magazine* 1, no. 3 (1910): 160–62.

Roberts, Isaac Phillips. *The Fertility of the Land: A Summary Sketch of the Relationship of Farm-Practice to the Maintaining and Increasing of the Productivity of the Soil.* New York: Macmillan, 1897.

Robichaud, Andrew A. *Animal City: The Domestication of America.* Cambridge, MA: Harvard University Press, 2019.

Rogers, R. H. "Purdue's Experimental Farms." *Purdue Agriculturalist* 17, no. 1 (1922): 7, 18.

Romo, David Dorada. *Ringside Seat to a Revolution: An Underground Cultural History of El Paso and Juárez, 1893–1923.* El Paso, TX: Cinco Puntos Press, 2005.

Roosevelt, Theodore. *The Letters of Theodore Roosevelt.* Cambridge, MA: Harvard University Press, 1951-.

Rosenberg, Gabriel N. *The 4-H Harvest: Sexuality and the State in Rural America.* Philadelphia: University of Pennsylvania Press, 2016.

———. "No Scrubs: Livestock Breeding, Eugenics, and the State in the Early Twentieth-Century United States." *Journal of American History* 107, no. 2 (2020): 362–87.

———. "The Trial of the Scrub Sire: Animal Gender and Eugenic Logics in the USDA's Better Sires—Better Stock Campaign, 1919–1940." Paper presented in the Department of History, University of Georgia, Athens, GA, 2017.

Roundtree, Charles. "Why I Breed Tunis Sheep." *Report of the Twenty-Sixth Annual Meeting of the Indiana Wool Growers' Association* (1901): 194–96.

Rowley, William D. *U.S. Forest Service Grazing and Rangelands: A History.* College Station: Texas A&M University Press, 1985.

"The Rural Science Series." *Cornell Daily Sun* 16, no. 20 (1895): [1].

Russell, Charles Edward. *The Greatest Trust in the World*. New York: Ridgway-Thayer, 1905.

Russell, Edmund. *Evolutionary History: Uniting History and Biology to Understand Life on Earth*. New York: Cambridge University Press, 2011.

Russell, Nicholas. *Like Engend'ring Like: Heredity and Animal Breeding in Modern England*. New York: Cambridge University Press, 1986.

Rydell, Robert W. *All the World's a Fair: Vision of Empire at American International Expositions, 1876–1916*. Chicago: University of Chicago Press, 1984.

———. *World of Fairs: The Century-of-Progress Expositions*. Chicago: University of Chicago Press, 1993.

Saddle & Sirloin Portrait Collection Guidebook. Louisville, KY: Merrick, 2020.

"Saddle & Sirloin Club Portrait Collection Historical Overview." North American Livestock Exposition, https://naile.s3.amazonaws.com/2021/06/2-History-SS.pdf.

Sanders, Alvin H. *At the Sign of the Stock Yard Inn*. Chicago: Breeder's Gazette Print, 1915.

———. "The Golden Age of Shorthorns." *Shorthorn in America* 4, no. 10 (1919): 10.

———. *A History of Aberdeen-Angus Cattle*. Chicago: New Breeder's Gazette, 1928.

———. *Red White and Roan*. Chicago: American Shorthorn Breeders' Association, 1936.

———. *Short-Horn Cattle: A Series of Historical Sketches, Memoirs, and Records of the Breed and Its Development in the United States and Canada*. Chicago: Sanders, 1900.

———. *Shorthorn Cattle: A Series of Historical Sketches, Memoirs, and Records of the Breed and Its Development in the United States and Canada*. Chicago: Sanders, 1916.

———. *The Story of the Herefords*. Chicago: Breeder's Gazette, 1914.

———. *The Story of the International from Its Inception in 1900 to the Show of 1941*. Chicago: International Live Stock Exposition, 1942.

———. "When the Show-Ring Hurts the Breeds: Disaster Follow 'Plugging' and Overfitting Aged Animals." *Wallaces' Farmer*, May 17, 1930, 1, 26.

Sanders, Elizabeth. *Roots of Reform: Farmers, Workers, and the American State, 1877–1917*. Chicago: University of Chicago Press, 1999.

Schanbacher, William D. *The Politics of Food: The Global Conflict between Food Security and Food Sovereignty*. Santa Barbara, CA: Praeger, 2010.

Schlebecker, John T. *Cattle Raising on the Plains, 1900–1961*. Lincoln: University of Nebraska Press, 1963.

Schrepfer, Susan R., and Philip Scranton, eds. *Industrializing Organisms: Introducing Evolutionary History*. New York: Routledge, 2004.

Schulman, Bruce J. *From Cotton Belt to Sunbelt: Federal Policy, Economic Development, & the Transformation of the South, 1938–1980*. Durham, NC: Duke University Press, 1994.

"Scrub Bull to Be Put On Trial! Why Not A Scrub Boar?" *Berkshire World* (1922): 21.

"Selling Pedigreed Stock." *Wallaces' Farmer*, March 6, 1908, 5.

"A Serious Condition." *National Provisioner* 26, no. 17 (1902): 19.

Shaw, E. L., and L. L. Heller. "Domestic Breeds of Sheep." *Bulletin of the United States Department of Agriculture* 94 (1914): 1–59.

Shaw, Thomas. *Animal Breeding*. New York: Orange Judd, 1901.

———. *The Study of Breeds in America: Cattle, Sheep, and Swine*. New York: Orange Judd, 1912.

Sheets, E. W., and M. A. R. Kelley. "Beef-Cattle Barns." *United States Department of Agriculture Farmers' Bulletin* 1350 (1923): 1–16.

Shepperd, J. H. "Breeding for and on the Range." *Proceedings of the First Annual Meeting of the American Breeders' Association* 1 (1905): 88–92.

——. *Livestock Judging Contests*. Fargo, ND: Agricultural Experiment Station North Dakota Agricultural College, 1922.

"The Show of Carlots of Fat and Feeder Cattle." *Breeder's Gazette* 82, no. 24 (1922): 841–42.

Sixth Annual Report of the Chicago Junction Railways and Union Stock Yards Company. Chicago: Chicago Junction Railways and Union Stock Yards Company, 1896.

Skaggs, Jimmy M. *Prime Cut: Livestock Raising and Meatpacking in the United States, 1607–1983*. College Station: Texas A&M University Press, 1986.

"A Sketch of President Guilliams and His Tunis Sheep." *Wool Markets and Sheep* 12, no. 4 (1901): 6.

Skinner, W. E. "Lifting the Lid." *Shepherd's Criterion* 15, no. 12 (1905): 11.

——. "Stock Sales at the International." *Wool Markets and Sheep* 11, no. 24 (1901): 28.

Slayton, Robert A. *Back of the Yards: The Making of a Local Democracy*. Chicago: University of Chicago Press, 1986.

Smil, Vaclav. *Enriching the Earth: Fritz Haber, Carl Bosch, and the Transformation of World Food Production*. Cambridge, MA: MIT Press, 2001.

Smith, Carl. *The Plan of Chicago: Daniel Burnham and the Remaking of the American City*. Chicago: University of Chicago Press, 2006.

Smith, Howard R. *Profitable Stock Feeding: A Book for the Farmer*. Chicago: Regan, 1906.

Smith, Jason Scott. *Building New Deal Liberalism: The Political Economy of Public Works, 1933–1956*. New York: Cambridge University Press, 2006.

"Smith's Standard Breeding Crate." *Berkshire World and Cornbelt Stockman* 2, no. 2 (1910): 20.

Soluri, John. *Banana Cultures: Agriculture, Consumption, and Environmental Change in Honduras and the United States*. Austin: University of Texas Press, 2005.

Sotham, T. F. B. "Building Meat on the Beef Model." *National Provisioner* 26, no. 3 (1902): 23.

——. "The Potency of Hereford Blood." In *Proceedings of the Third Annual Convention of the National Live Stock Association*, 343–48. Denver, CO: Smith-Brooks, 1900.

Soule, Andrew M. "What Bull Should Be Used?" *Breeder's Gazette* (1900): 71.

Specht, Joshua. *Red Meat Republic: A Hoof-to-Table History of How Beef Changed America*. Princeton, NJ: Princeton University Press, 2019.

——. "The Rise, Fall, and Rebirth of the Texas Longhorn: An Evolutionary History." *Environmental History* 21 (2016): 343–63.

Spoor, J. A. "Tells of Great Year." In *"Our Year Book": Telling Tables of the Livestock Trade for the Year 1902*, 11–12. Chicago: Chicago Daily Drovers Journal, 1903.

"Spotlight Stars of the University of California." *University of California Journal of Agriculture* 4, no. 5 (1917): 160–61.

"Steady Progress Shown in Campaign for Better Sires." *Weekly News Letter of the U.S Department of Agriculture* 7, no. 41 (1920): 1.

Stern, Alexandra Minna. *Eugenic Nation: Faults and Frontiers of Better Breeding in Modern America*. 2nd ed. Oakland, CA: University of California Press, 2016.

Stewart, Elliot W. *Feeding Animals: A Practical Work upon the Laws of Animal Growth.* 4th ed. Buffalo, NY: Baker, Jones, 1888.

Stewart, Henry. *The Domestic Sheep: Its Culture and General Management.* Chicago: American Sheep Breeder Press, 1900.

"Stock Breeders' Exposition." *New York Times*, November 25, 1899.

"Stock Show to Be Great." *Chicago Daily Tribune*, August 4, 1900.

Stoll, Steven. *Larding the Lean Earth: Soil and Society in Nineteenth-Century America.* New York: Hill and Wang, 2002.

Striffler, Steven. *Chicken: The Dangerous Transformation of America's Favorite Food.* New Haven, CT: Yale University Press, 2005.

Strom, Claire. *Making Catfish Bait out of Government Boys: The Fight against Cattle Ticks and the Transformation of the Yeoman South.* Athens: University of Georgia Press, 2009.

"Suggestions for Keeping Boys on Farms." *Breeder's Gazette* 53, no. 4 (1908): 175.

Swift and Company. *The Meat Packing Industry in America.* Chicago: Swift and Company, 1920.

Swift, Louis F., and Arthur Van Vlissingen Jr. *The Yankee of the Yards.* Chicago: A. W. Shaw, 1927.

"System of Robbing the Soil." *Prairie Farmer* 77, no. 38 (1905): 3.

"Tendency Shown toward Use of Purebred Female Stock." *Weekly News Letter of the U.S. Department of Agriculture* 8, no. 26 (1921): 10.

Terrill, Clair E. "Fifty Years of Progress in Sheep Breeding." *Journal of Animal Science* 17, no. 4 (1958): 944–59.

Third Annual Report of the Chicago Junction Railways and Union Stock Yards Company. Chicago: Chicago Junction Railways and Union Stock Yards Company, 1893.

Tomson, Frank D. "Shorthorn Excellence." *Shorthorn World* 4, no. 1 (1919): 11.

Tormey, J. L. "International Just Out of Its Teens: A Running Review of Some of the High Lights in the Greatest Live Stock Show in the World." *Shorthorn World* 4, no. 18 (1919): 15, 139–40, 149.

Toole, Wade. "Development of Our Modern Beef Type." *Shorthorn World* 4, no. 23 (1920): no. 9-11.

"To Maintain Soil Fertility." *Wallaces' Farmer*, March 8, 1901, 298.

"Transfer of Pedigree." *Wallaces' Farmer*, May 6, 1910, 2.

True, Alfred Charles. *A History of Agricultural Education in the United States, 1785–1925.* Washington, DC: Government Printing Office, 1929.

Tucker, Richard P. *Insatiable Appetite: The United States and the Ecological Degradation of the Tropical World.* Lanham, MD: Rowman and Littlefield, 2007.

"20th Anniversary of International." *Shorthorn World* 4, no. 14 (1919): 80.

"Two Methods of Farming." *Farmers' Review* 35, no. 27 (1904): 469.

Unfer, Louis. "Swift and Company: The Development of the Packing Industry 1875 to 1912." PhD diss., University of Illinois, 1951.

"Union Stockyards." *Chicago Daily Tribune*, October 10, 1904.

"Unite for a Big Stock Show." *Chicago Daily Tribune*, November 25, 1899.

United States Department of Agriculture. "Statistics of Agriculture: Introduction." In *1900 Census Publications: Farm, Livestock, and Animal Products* 5, pt. 1. Washington, DC: United States Census Office, 1902.

"The United States Meat Industry at a Glance." North American Meat Institute, https://www.meatinstitute.org/index.php?ht=d/sp/i/47465/pid/47465.

"United States v. One Hundred and Ninety-Six Mares." *Federal Reporter* 5 (1897): 139.

"Urban vs. Rural Population." *Wallaces' Farmer*, December 28, 1906, 554.

"Vaccine Protects Feeder Lambs." *Wallaces' Farmer*, October 7, 1950, 62.

"Value of Pedigree to the Stockman." *Wallaces' Farmer*, December 5, 1902, 1608.

Van Norman, George B. "Live Stock Exchanges and Their Relation to the Producer." In *Proceedings of the Fourth Annual Convention of National Live Stock Association* 384–87. Denver, CO: Smith-Brooks, 1901.

Van Wagenen, Bleecker. "Preliminary Report of the Committee of the Eugenics Section of the American Breeders' Association to Study and to Report on the Best Practical Means for Cutting Off the Defective Germ-Plasm in the Human Population." In *Problems in Eugenics: Papers Communicated to the First International Eugenics Congress*, 460–79. London: Knight, 1912.

Vaughan, Henry William. "A Picture of the Live Stock Industry." *Marketing Live Stock* 1 (1922): 1–36.

———. *Types and Market Classes of Livestock*. Columbus, OH: R. G. Adams, 1915.

Veit, Helen Zoe. *Modern Food, Moral Food: Self-Control, Science, and the Rise of Modern American Eating in the Early Twentieth Century*. Chapel Hill: University of North Carolina Press, 2013.

"Vet Advises Treating Lambs for Worms." *Farmers' Weekly Review* 31, no. 37 (1952): 2.

Wade, Louise Carroll. *Chicago's Pride: The Stockyards, Packingtown, and the Environs in the Nineteenth Century*. Urbana: University of Illinois Press, 1987.

"Ward, Unclean, Kills Babes." *Chicago Daily Tribune*, August 4, 1910.

Warren, G. F. *Elements of Agriculture*. New York: Macmillan, 1909.

Warwick, E. J. "Fifty Years of Progress in Breeding Beef Cattle." *Journal of Animal Science* 17, no. 4 (1958): 922–43.

"Watch Your Feeder Lambs, They May Eat Themselves to Death." *Farmers' Weekly Review* 27, no. 35 (1948): 4.

"Weber, Arthur D." North American Livestock Exposition. https://naile.s3.amazonaws.com/2021/06/8-Portraits-P-Z.pdf.

Wentworth, Edward N. *A Biographical Catalog of the Portrait Gallery of the Saddle and Sirloin Club*. Chicago: Union Stock Yards, 1920.

———. *Progressive Beef Cattle Raising*. Chicago: Armour's Bureau of Agricultural Research and Economics, 1920.

———. *Progressive Hog Raising*. Chicago: Armour's Bureau of Agricultural Research and Economics, 1922.

———. *Types and Market Classes of Livestock*. Columbus, OH: R. G. Adams, 1915.

"What Farmers of the Middle States Must Do." *Breeder's Gazette* 1, no. 20 (1882): 494.

"What Is Baby Beef?" *Breeder's Gazette* 70, no. 1 (1916): 26–27.

White, Richard. "Animals and Enterprise." In *The Oxford History of the American West*. Eds. Clyde A. Milner II, Carol A. O'Connor, and Martha A. Sandweiss, 237–73. New York: Oxford University Press, 1994.

———. *Railroaded: The Transcontinentals and the Making of Modern America*. New York: Norton, 2011.

Whitford, Frederick. *For the Good of the Farmer: A Biography of John Harrison Skinner*. West Lafayette, IN: Purdue University Press, 2013.

Whitford, Frederick, and Andrew G. Martin. *The Grand Old Man of Purdue University and Indiana Agriculture: A Biography of William Carroll Latta*. West Lafayette, IN: Purdue University Press, 2005.

Whitford, Fredrick, Andrew G. Martin, and Phyllis Mattheis. *The Queen of American Agriculture: A Biography of Virginia Claypool Meredith*. West Lafayette, IN: Purdue University Press, 2008.

Whitson, Jay. "Baby Beefmaking in the Cornbelt." *Breeder's Gazette* 79, no. 14, (1921): 633–34.

———. "Cattle Feeding Methods Changing?" *Breeder's Gazette* 80, no. 13 (1921): 437–38.

Willham, R. L. "Genetic Improvement of Beef Cattle in the United States: Cattle, People and Their Interaction." *Journal of Animal Science* 54, no. 3 (1982): 659–66.

Williams, R. H. "Eliminating Hazards in the Range-Cattle Business." *Producer* 3, no. 3 (1921): 5–8.

Wing, DeWitt C. "The Present Situation in the Live Stock World." *Berkshire World and Cornbelt Stockman*, February 1925, 20.

Winkleman, Henry. "Problems of American Agriculture." *Breeders' Gazette* 34, no. 6 (1898): 104.

"With a Long Pull and Strong Pull They'll Get Odors from 'Bubbly.'" *Chicago Daily Tribune*, October 30, 1910.

"With Sanders in the Saddle and Sirloin Hall." *Clay, Robinson and Company Livestock Report*, June 22, 1916, 10–11.

"Women as Animal Husbandmen." *Breeder's Gazette* 64, no. 3 (1918): 85.

"The Woman Movement and Eugenics." *American Breeders Magazine* 2, no. 3 (1911): 225–28.

Woods, Rebecca J.H. *The Herds Shot Round the World: Native Breeds and the British Empire, 1800–1900*. Chapel Hill: University of North Carolina Press, 2017.

"The World's Greatest Stock Show." *Breeder's Gazette* 76, no. 21 (1919): 1094d.

"Yankees Would Swap Rhine for Bubbly Creek." *Chicago Daily Tribune*, February 14, 1919.

Yearbook of the United States Department of Agriculture, 1921. Washington, DC: Government Printing Office, 1922.

Yearbook of the United States Department of Agriculture, 1936. Washington, DC: Government Printing Office, 1936.